AF008241

# PCR TROUBLESHOOTING

# THE ESSENTIAL GUIDE

by Michael L. Altshuler

Mechnikov Institute of Vaccine and Sera
Department of Microbiology,
Maly Kazenny pereulok, 5A
105064 Moscow
Russia

Cover picture adapted from Broude et al. (2004) High multiplexity PCR based on PCR suppression. In DNA Amplification: Current Technologies and Applications, V.V. Demidov and N.E. Broude, ed. (Wymondham, UK: Horizon Bioscience), pp. 61-76.

 Caister Academic Press

Copyright © 2006
Caister Academic Press
32 Hewitts Lane
Wymondham
Norfolk NR18 0JA
U.K.

www.caister.com

**British Library Cataloguing-in-Publication Data**

A catalogue record for this book is available from the British Library

ISBN-10: 1-904455-07-7 (paperback)
ISBN-13: 978-1-904455-07-3 (paperback)

ISBN-10: 1-904455-08-5 (ebook)
ISBN-13: 978-1-904455-08-0 (ebook)

Description or mention of instrumentation, software, or other products in this book does not imply endorsement by the author or publisher. The author and publisher do not assume responsibility for the validity of any products or procedures mentioned or described in this book or for the consequences of their use.

*All rights reserved. No part of this publication may be reproduced, stored in a retrieval system, or transmitted, in any form or by any means, electronic, mechanical, photocopying, recording or otherwise, without the prior permission of the publisher. No claim to original U.S. Government works.*

*Printed and bound in Great Britain*

# TABLE OF CONTENTS

**PREFACE** .................................................................9

**INTRODUCTION** ....................................................11

**APPEARANCES**......................................................13
- **Unsatisfactory results of PCR** ............................13

**CAUSES AND ACTIONS** .......................................14
- **Do not confront the problem** .............................14

- **The nature of PCR and its pathology** ................15

- **Inadequate concentrations of ingredients** ........18
    1. The template ..................................................18
    2. Inadequate deoxynucleotidetriphosphates......20
    3. Inadequate primers ........................................20
    4. Inadequate Taq ..............................................21
    5. Inadequate $Mg^{2+}$ ...........................................21
    6. Suboptimal KCl concentration in the PCR buffer or the whole of the PCR buffer....................21

- **Inadequate quality of ingredients** ......................23
    1. DNA template..................................................23
        ➢ Conversion of a DNA solution into a solid body ......23
        ➢ PCR inhibitors..........................................23
        ➢ Degraded template .................................23
        ➢ Verification of the purity of the template DNA by optical means............................24
    2. Poor water .......................................................24
    3. Deoxynucleotidetriphosphates ........................24

- 4. Poor primers .................................................................25
  - ➤ Primers may not be good in practice even if they are good in design .........................................25
  - ➤ For primer selection use only dedicated software ..............................................................25
  - ➤ Difference in $T_m$ balanced by different primer concentrations ...........................................26
  - ➤ Inefficient priming ...............................................26
- 5. Inadequate $MgCl_2$ ......................................................26
- 6. Poor buffer ...................................................................27
- 7. Poor Taq .....................................................................27
- 8. The presence of PCR inhibitors ...................................27
- 9. Substances that <u>do not</u> inhibit PCR ............................28

- **Inadequate storage of ingredients for the PCR reaction** ..29
  - 1. The treacherous refrigerators ....................................29
  - 2. Templates, dNTPs, primers .......................................29
  - 3. Taq polymerase .......................................................31
  - 4. $MgCl_2$ ......................................................................31
  - 5. Buffer ......................................................................31

- **The thermocycler** ........................................................32
  - 1. Conductivity of heat puts a limit to the mix composition ..............................................32
  - 2. The ramp ...............................................................33
  - 3. Time and temperature ..............................................34
  - 4. Dusty, greasy, or fluffy tube wells .............................36
  - 5. Rapid evaporation ...................................................36
  - 6. Suboptimal performance of the thermocycler or its particular wells ....................................................36
  - 7. The tubes are poorly pressed down into the wells or they have got deformed ......................................37

- **Faulty target selection** ...................................................... 38

- **Incomplete DNA denaturation and dispersal** ................ 40
    1. Template DNA ............................................................... 40
    2. PCR fragments ............................................................... 41
    3. Hairpins .......................................................................... 42

- **The Taq enzyme** ...................................................................... 43
    1. Hurdles for Taq polymerase .......................................... 43
        ➢ Stable hairpins in the template strand .................. 43
        ➢ AT-rich areas ........................................................... 43
        ➢ GC-rich areas ........................................................... 43
        ➢ Alternating GC/AT-rich regions ........................... 43
    2. Hot start ........................................................................... 44
        ➢ The improved hot start ............................................ 44
        ➢ The role of the reaction volume in the
          quasi hot start ............................................................ 45
    3. Nonspecific binding of Taq to DNA ............................ 46

- **Incomplete primer elongation
  or premature termination of DNA synthesis** ................. 48
    1. Under-elongation of primers in the late PCR ............. 48
    2. Premature termination of the DNA synthesis ............ 50

- **Cosolvents or additives or enhancers** ............................... 51
    1. Helix-destabilizers ......................................................... 51
    2. Helix-stabilizers ............................................................. 52
    3. Substances that neutralise the PCR inhibitors ........... 53
    4. PCR enhancers with poorly understood
       mechanism of action ....................................................... 53

- **Approaching the limit of the PCR sensitivity** ................ 54

- **Agarose gel electrophoresis** ......................................... 57
    1. The band diffuses and disappears ................................ 57
    2. Short fragments of uneven length migrate
       down the gel without any separation ........................... 57
    3. The band is invisible ................................................ 57
    4. The significance and the insignificance of the salt
       concentration in the compared samples ....................... 57
    5. Bands smear due to their fast movement or the
       DNA overload ......................................................... 58
    6. Dirty gel support .................................................... 58
    7. Dried well ............................................................. 58
    8. DNA sticking in the gel well caused by
       inappropriate gel density ......................................... 58

- **Causes for specific nonspecifics
   and the false contamination** ........................................ 59
    1. Chimeras .............................................................. 59
    2. Allele dropout ....................................................... 59
    3. Heteroduplexes ...................................................... 60
    4. Primer multimers ................................................... 61
    5. Low resolution electrophoresis resulting in
       imprecise idea of the correct band position ................. 61
    6. Coincidence or the devil ......................................... 61

- **Mineral oil and wax** .................................................. 62
    1. Mineral oil ........................................................... 62
    2. Wax, paraffin or vaseline ........................................ 62

- **Primer-dimers and primer multimers** ........................... 63

- **Short PCR fragments versus long PCR fragments** ........ 67

- **Avoiding accidents** ................................................... 68

**CONCLUSIONS** .................................................................. 69
- **A few words to the novice** ........................................... 69

- **A few words to a PCR adept** ....................................... 69

**GLOSSARY** ..................................................................... 71

**INDEX** ............................................................................ 77

# Other Books of Interest

- Real-Time PCR: An Essential Guide
- DNA Amplification: Current Technologies and Applications
- Microbial Bionanotechnology
- Molecular Diagnostics: Current Technology and Applications
- DNA Microarrays: Current Applications
- Computational Biology: Current Methods
- SAGE: Current Technologies and Applications
- Protein Expression Technologies: Current Status and Future Trends
- Computational Genomics: Theory and Application
- The Internet for Cell and Molecular Biologists (2nd Edition)
- Peptide Nucleic Acids: Protocols and Applications (2nd Edition)

**Full details of all these books at:**
**http://www.horizonpress.com**

# PREFACE

"A broad road full of wonderful prospects is about to open up before you," my boss said hinting at my impending firing, after taking a look at an empty gel. "A clear example of complete professional degeneracy," he greeted me the next time. He is a fine man, I like him very much. His greetings, congratulations and wishes of good fortune outside his laboratory have goaded me into an idea to compile a PCR troubleshooting guide of a novel type accompanied by a brief consideration of ways to obviate rather than overcome the PCR problems. If you fail at PCR, consult this book. Then try to figure out the most probable cause of your failure.

I wouldn't let this guide slip into a full-scale manual, so inevitably it has had to be oriented to a reader with some experience in PCR. If you are a PCR greenhorn, keep several manuals close at hand while reading it. Although there is a considerable overlap with other troubleshooting guides, some of the most obvious advice, the kind that you generally remember and is invariably mentioned in other textbooks and guides, has been deliberately omitted from my book (detailed account of the hot start, the trick of touchdown, commonly accepted rules for primer selection, the significance of contamination, etc.).

The advice that you will find in these pages is the sort of advice that is not usually found elsewhere and that is often the most useful.

Good luck in PCR and elsewhere!

I wish to express my deep gratitude to the laboratory heads, Prof. A.I. Gluckov and Prof. L.P. Blinkova, who have permitted me to write this article without stopping my salary. The boss mentioned above should not be identified with either of them.

Michael L. Altshuler
Moscow, March 2006

# INTRODUCTION

If common (conventional) PCR is allowed to be considered a lore and a branch of applied science, rather than a single and simple method, it should be amenable to subdivision into several specific areas. In this book, I have made an attempt to represent PCR as a set of distinct, albeit overlapping, issues. For example, the issue of denaturation which, in turn, breaks down into three sub-sections: (1) the problem of denaturing the template DNA at different levels of its fragmentation, (2) conditions for denaturing PCR-fragments, and (3) denaturation of stable hairpins. There is the issue of the correct primer selection, the problems of PCR inhibitors and PCR artifacts (such as chimeras and allele dropout), and the intriguing area of cosolvents. Most troublesome seems the issue of the occurrence of the incomplete primer elongation leading to a miscellany of unlucky PCR outcomes including several kinds of grave artifacts and PCR products unsuitable for the purpose for which they were intended (e.g., sequencing, cloning, mutation detection by SSCP or allele-specific amplification). This collection of undesirable phenomena probably includes the formation of multi-chained molecular aggregates having altered electrophoretic mobility and readily misinterpreted as the common nonspecifics.

Theory is one thing; practice is another. I have noticed long ago that when a real adept in PCR is considering the causes of an unsuccessful experiment she/he is prone to overlook some of the most evident. It is a psychology rather than biology, but that does not make things better. Thus, I believe that a meticulous written reminder of the causes is in demand together with a firm rule to consult it every time you get a failure, no matter how sure you are of the cause of the failure. I sincerely hope that this book will fulfill the role of "reminder".

In this book I consider "APPEARANCES" and "CAUSES AND ACTIONS". **APPEARANCES** is what is actually seen in the gel: nothing, additional bands, smear, etc. **CAUSES** is what the appearances result from and **ACTIONS** are the steps that should be taken to alter a particular appearance. "CAUSES AND ACTIONS" can be considered in two categories: 1) indirect ways to correct an appearance; 2) molecular phenomena reflecting the complexity of events comprising the nature of the PCR process.

This book is a collection of errors; being that, it represents an attempt to circumscribe the field of the basic PCR and define within this boundary a set of distinct areas. Different issues are often interconnected, so the PCR lore, in fact, resembles a network. This is why the book abounds in cross-references.

The guide is intended to be read as a book or to be used as an electronic document in PDF format. Each **APPEARANCE** and some **CAUSES AND ACTIONS** are assigned a letter code for brevity and to facilitate the "search" feature of a PDF reader such as Adobe Acrobat. Codes for *appearances* are presented in ***UPPERCASE BOLD ITALIC***; codes for *causes and actions* in **UPPERCASE BOLD**. The codes for appearances are to be used for troubleshooting. The codes for causes and actions have been introduced for the ease of crossreferencing. Codes for appearances are described in the following page. In addition, definitions of all codes can be found in the glossary at the back of the book. If you have the printed form of this book, use the page index for codes after the glossary at the back of the book. If you have the electronic version, search for a code or for a keyword relevant to your work, e.g., "hairpins", "low copy number", "long PCR products", etc.

# APPEARANCES

## *UNSATISFACTORY RESULTS OF PCR*

Unsatisfactory results of a polymerase chain reaction as they appear in the visual examination of an agarose gel stained with ethidium bromide can be as follows:

- Blank gel (***BLANK***) – no bands, with the possible exception of the bands at its bottom containing associated primers.

- Ladder (***LADD***) – one or several bands instead of, or in addition to, the required band; low molecular weight entities comprising primer aggregates at the very bottom of the gel are not considered as components of ladder.

- Smear (***SMEA***) – a smoothly stained area in the lane having no stepwise character of the ladder.

- Specific nonspecifics (***SPNO***) – unsatisfactory result may *appear* as a satisfactory one. A single band in the lane in the expected position in fact is not identical to the target that has had to be amplified.

- The proper band occupies improper position in the gel (***PBOI***).

- Untidy appearance of the gel with stained blots, irregular patches, dots, etc. (***UNTIDY***).

- A stained image of the well slowly migrating down the lane (***GHOS***).

- DNA remains in the well and won't enter the gel (***HMWG***).

- The trivial contamination or false contamination (***TRIV***).

# CAUSES AND ACTIONS

## *DO NOT CONFRONT THE PROBLEM – TAKE A WALK AROUND IT*

- If you repeatedly get the ***LADD*** appearance, abandon further attempts at optimization. Do the nested PCR. Alternatively, reamplify the desired band from agarose or an empty place in the lane which the band must have occupied. Use a protocol described by E. Rybicki in http://www.mcb.uct.ac.za/manual/corepcr.htm or just scrape the band with a toothpick and rinse it in the PCR tube. Mind **BTIP**.

- If you get ***BLANK***, do not bother to search for a cause. Dilute the amplification product 100-fold and use it as a template for another amplification. Mind **RQWN**.

- In case of ***BLANK***, there is usually a reason to presume that any ingredient of the PCR mix is not OK. The straightforward way to verify that is to replace the ingredients one by one or all at once with those taken from new batches. There is a simpler or cheaper means: repeat the PCR at low annealing temperature. If there is some specific or nonspecific product, then all the substances are probably good.

- If you are unable to optimize PCR (***BLANK***, ***LADD***, ***SMEA***), and you think that the reason may be an inadequacy of one of the primers, try the following course of action: add to the mix just one primer that is supposed to be good and do a 35 cycle PCR; then add the other primer and a new aliquot of the Taq enzyme and carry out the PCR under conditions supposed to be appropriate. This approach has worked excellently in my own hands after a long and frustrating period of unsuccessful optimization. In theory, such an approach could also be helpful in the situations when primers are capable of forming aggregates or fold into hairpins (see **PRDI**). Use of only one primer, if it is incapable of interacting with itself, could raise the target concentration 10-30 fold and, upon the subsequent addition of the other primer, can be expected to shift the balance of the reaction in the right direction (**SBRD**).

- (***LADD***) When the DNA samples which have to be compared by electrophoresis are dissolved in the media of the different salt concentration, do not bring them to the even salt concentration but do as described under **HREN**.

## *THE NATURE OF THE PCR AND ITS PATHOLOGY*

The following list is aimed at creating an integrated and concise view of the fundamental aspects of PCR and of the problems that can occur. The actual content of this book is more voluminous and elaborate yet less comprehensive, being biased towards the dark, fine and readily overlooked problems of PCR.

- PCR kinetics and PCR efficiency; loss of quantitative effect after the end of the logarithmic phase and the pressing demand for the quantitative PCR in medicine.

- The computer design of a PCR reaction: selection of the target; selection of the primers; primer verification; inspection of the inter-primer region for runs of AT- or GC-rich sequence as well as for the stable hairpins.

- Concentrations of ingredients.

- Quality of ingredients including the problem of the PCR inhibitors.

- DNA denaturation and dispersal.
    1. Template DNA.
    2. PCR fragments.
    3. Hairpins.

- Incomplete primer elongation (premature termination of DNA synthesis).

- The Taq enzyme.
    1. Hurdles for Taq polymerase.
        - Stable hairpins in the template strand.
        - AT-rich areas.
        - .....more.......
    2. The hot start
    3. Nonspecific binding of Taq to DNA.

- Primer selection.

- The thermocycler.
    1. The temperature regulation in a thermocycler.
    2. The ramp.
    3. Time and temperature for the three components of a regular amplification cycle: annealing, synthesis, denaturation. Touchdown.
    4. Choice of the number of cycles.

- Agarose gel electrophoresis.
    1. The band diffuses and disappears after a long run in thin or low density gel.
    2. ......more......

- Cosolvents.

- Approaching the limit of the PCR sensitivity.

- Spurious results.
    1. Heteroduplexes.
    2. Allele dropout.
    3. ......more......

- Mineral oil and wax.

- Primer oligomers and multimers.

- The problem of contamination.

# INADEQUATE CONCENTRATIONS OF INGREDIENTS

## 1. THE TEMPLATE

Total DNA added to a PCR tube comprises the target sequence to be amplified and the nontarget DNA which could be called "DNA burden". Different proportions of the target to nontarget DNA lead to different outcomes (Table 1).

| Table 1. PCR with different ratios of the total/target DNA | | |
|---|---|---|
| Amount of the total DNA | Copy number of the target DNA | Possible outcome and advice |
| excessive | excessive<br>medium<br>small | ***BLANK, LADD, SMEAR*[1)]** |
| medium[4)]<br>(1ng – 0.5µg per 50µl reaction) | excessive[2)] | little or no sense in doing PCR |
| | medium ($10^3 - 10^8$ target molecules per 50µl reaction) | *LADD*[3), 6)] |
| | small (down to one copy) | a challenging task, see **ALPS** |
| small[7)] | excessive[2)] | little or no sense in doing PCR |
| | medium ($10^3 - 10^8$ target molecules) | *LADD*[3)] |
| | small (down to one copy) | see note #5 below the table |

1) The packed DNA in a closed space of the tube can be expected to lead to false priming and probably to poor DNA synthesis because of the obstructed diffusion of large Taq polymerase molecules (***LADD***, ***BLANK***, ***SMEAR***, ***HMWG***). The ***SMEA*** would represent the original DNA rather than the PCR result. The immense excess of the long strands of the viral DNA is likely to make visible the whole of the initial DNA input sticking in the gel wells (***HMWG***).

2) The input of greater than $10^{10}$ copies of a 500bp PCR fragment (5ng) could be visible in the agarose gel stained with ethidium bromide.

3) The target concentration should be balanced with the number of cycles. The elevated concentration of the target combined with the usual number of cycles will provide a reason for the appearance of ***LADD***, because the saturating concentration of the fragment resulting from the target will be gained the sooner and the cycling continued thereafter will favor accelerated accumulation of the nonspecific products. The accumulation of the nonspecific products is readily observed even in reamplification, when a high initial concentration of the PCR fragment is accompanied by a high number of cycles (**RQWN**).

4) The boundary between "medium" and "small" quantity of the total DNA is drawn on the ground of the evidence that handling tiny amounts of DNA often requires special precautions. Furthermore, with the human genome which seems to be the most common object for PCR the indicated range of 1ng-0.5µg usually provides satisfactory results [O. Henegariu http://www.info.med.yale.edu/genetics/ward/tavi/p09.html , O. Zymnik, pers. comm.].

5) A challenging task when the ratio of the target copy number to the burden DNA is as that in the human genome (1 to $6 \cdot 10^6$ for 500bp PCR fragment) or less (see **ALPS**); much easier, if the ratio is about 1 to $10^4$ as in the *Escherichia coli* genome or 1 to 1 as in reamplification.

6) "Low primer, target, Taq, and nucleotide concentrations are to be favoured as these generally ensure cleaner product and lower background" [E. Rybicki, http://www.mcb.uct.ac.za/manual/pcrcond.htm] (***LADD***, ***SMEAR***) (**BBBB**).

7) When the total amount of the input DNA is extremely small, there is an increased likelihood of its loss owing to any conceivable cause (clotting, adsorption, chemical or enzymatic degradation). Furthermore, if the total amount of the added DNA *must be* very small, there emerges a risk of introducing additional DNA from impurities on anything that can come in contact with the DNA solution (**ADNA**). In this respect, both the DNA diluent, the dust floating abroad in the air, exhalations and the flakes of your body should not be disregarded, as these can carry both the DNA and the DNA-degrading substances. Nucleases are recognized as the major agents

of DNA degradation. They are abundant on the surface of the human skin and can be present everywhere else. If the substances will withstand it, you can get rid of both the unwanted DNA and the nucleases by wholesale mild autoclaving the DNA diluent and everything that comes in regular, occasional, or accidental contact with the extremely diluted DNA solutions. Put your hands in autoclaved gloves, stick your hair under a surgeon cap, breathe through a gauze filter, wash the working space with an oxidating solution (6% $H_2O_2$), and remember that the UV irradiation of the working area should be considered only as an auxiliary measure, rather half-useless than quite useless (*BLANK*, a few contaminating amplicon molecules could lead to *SPNO*) (**HTHT**).

## 2. INADEQUATE DEOXYNUCLEOTIDETRI-PHOSPHATES

(*BLANK*, *LADD*, *SMEA*, *SPNO*)

The usual dNTP concentration is between 40-200μM EACH of the four. *Excessive dNTP* inhibit PCR (*BLANK*) [P. Frame, http://www.ufbi.ufl.edu/~rowland/protocols/pcr.htm]. However, some researchers trespass this range. M.S.B. Judo et al. have used 400 μM each dNTP [Nucl. Acids Res., 1998, 26:1819] (**EDNE**). See **PRST**, **BBBB**. *Deficient deoxynucleotidetriphosphates*. For longer PCR-fragments try a higher deoxynucleotidetriphosphate concentration (*BLANK*, *LADD*). Suboptimal concentration of nucleotides exacerbates **IPEL** (**APPT**).

## 3. INADEQUATE PRIMERS

The common and advisable primer concentration range is 0.1-1μM each primer. If the primers are not prone to making the dimers, the high primer concentration would result in *LADD*. If the primers form dimers, the raising of their *actual* concentration would likely not result in the raising of their *effective* concentration, since the more primers you add, the more dimers they generate. Eventually you will miss the desired band (*BLANK*) and get visible only the primer-derived oligomers [P. Frame, http://www.ufbi.ufl.edu/~rowland/protocols/pcr.htm] (**GVPT**). See **BBBB**.

Getting a visible band with short PCR fragments requires more molecules and that requires more primers. Suppose the fragment length being 100bp, the reaction volume 50μl, the input per gel well 10μl. Assuming that the clearly visible band must contain 50ng DNA, the final number of the PCR fragment molecules must be $2.5 \cdot 10^{12}$. The initial number of the primer molecules in the same volume at 1μM concentration of each primer will be $3 \cdot 10^{13}$ each primer, only 10 times higher and barely enough to finish the business (*BLANK*) (**MPRE**).

## 4. INADEQUATE TAQ

(*BLANK*, *LADD*, *SMEA*, *SPNO*).
Use approximately 1 unit of the enzyme for 25μl reaction (O. Henegariu, http://www.info.med.yale.edu/genetics/ward/tavi/p01.html). Suboptimal concentration of the Taq will exacerbate **IPEL**. Excessive Taq will result in *SMEAR* and the huge excess will result in *BLANK* (**KPPT**). See **BBBB**.

## 5. INADEQUATE Mg$^{2+}$

Causes *BLANK* or *LADD*, perhaps *SMEA*. The limits of the MgCl$_2$ concentration are usually 1-4mM [P. Frame, http://www.ufbi.ufl.edu/~rowland/protocols/pcr.htm]. Since dNTPs sequester Mg$^{2+}$, a gross change in the dNTP concentration would certainly require a change in the concentration of MgCl$_2$ (**PRST**). Also, a change in the KCl-based buffer concentration [O. Henegariu http://www.info.med.yale.edu/genetics/ward/tavi/p14.html] or any other component of the PCR mix may or may not require readjustment of the Mg$^{2+}$ concentration. (**READ**).

## 6. SUBOPTIMAL KCI CONCENTRATION IN THE PCR BUFFER OR THE WHOLE OF THE PCR BUFFER

(*BLANK*). See O. Henegariu at http://www.info.med.yale.edu/genetics/ward/tavi/p07.html and http://www.info.med.yale.edu/genetics/ward/tavi/p06.html (**SKCI**).

# INADEQUATE QUALITY OF INGREDIENTS

## 1. DNA TEMPLATE

### ➢ CONVERSION OF A DNA SOLUTION INTO A SOLID BODY

A solution of high molecular weight DNA (e.g., resulting from phenol/chloroform extraction) within an eppendorf occasionally turns into a transparent semi-solid body which gives off some watery DNA-free substance on its surface. The DNA from the body cannot be taken up by pipette. This inconspicuous transition from the liquid to the solid state can be overlooked and the liquid overlaying the body can be believed to be the DNA and added to a PCR reaction tube (***BLANK***).

### ➢ PCR INHIBITORS

The template can contain PCR inhibitors (***BLANK***). See P. Frame, http://www.ufbi.ufl.edu/~rowland/protocols/pcr.htm and **TEIN**. Inhibitors in the template is one of the most common causes of ***BLANK***. Dilute the template tenfold and repeat the PCR (**DILU**).

### ➢ DEGRADED TEMPLATE

(***BLANK, LADD***). *Verification of the integrity of the template DNA by means of electrophoresis.* If the DNA does not contain very short fragments resulting from degradation or excessive fragmentation, it will migrate down the agarose gel as a single tight band for a greater time/distance than the DNA which contains them. With the preparation obtained with phenol-chloroform extraction followed by ethanol precipitation the good DNA typically retains the single band appearance after 2 hours in 0.8% agarose.

If the RNA has not been removed, there are two additional bright bands in the lane. With the phenol-chloroform extraction, the DNA band is usually the uppermost (**KLMN**).

> ### VERIFICATION OF THE PURITY OF THE TEMPLATE DNA BY OPTICAL MEANS

(***BLANK***). A rough estimate of contamination of the DNA preparation with proteins can be obtained by the reading of $A_{260}/A_{280}$. In certain rare cases the result can be false. The $A_{260}/A_{280}$ value depends also on pH and the ionic strength. Both must be constant when comparisons are made. The sensitivity is greater at higher pH and low ionic strength [T. Fitzwater, BioTechniques Molecular Biology Techniques Forum, General Methods: http://molecularbiology.forums.biotechniques.com/forums/viewforum.php?f=5 (search for a response to the question by *Spudlab* posted on Thu Jun 26, 2003)]. In TE buffer at pH8.0 the absorbance of the protein-free and RNA-free DNA should be 1.8. As far as I know, if the DNA concentration is insufficient, there is no easy way to check its purity and estimate the degree of its integrity. Select a PCR-compatible method for DNA isolation and rely on the accuracy of your manipulations.

## 2. POOR WATER

Poorly deionized water causes ***BLANK***. Bidistilled water is insufficiently clean. Use tridistilled or cleaner water. If the water is contaminated with nucleases, it is ***BLANK***. Autoclaving can help to inactivate nucleases as well as some PCR inhibitors present in the water (**AQWW**).

## 3. DEOXYNUCLEOTIDETRIPHOSPHATES

(***BLANK***). See **KKIM, PYRO, IPMC**.

## 4. POOR PRIMERS

(**BLANK**, **LADD**, **SMEA**). *Several tips on primer selection.* The computer design of a PCR reaction usually implies four steps: (1) choice of the target sequence. See **FTSE**. (2) primer selection by a software followed by the critical and judicious visual evaluation of primers offered by the computer; (3) primer verification for its incidental similarity to repeats (in eukaryotes) using the *alu* library in BLAST – in addition to its filtering through a repeat library integrated into the primer selection software; (4) analysis of the internal (interprimer) region for the significantly long runs of AT or GC-rich sequence and for the potential for the formation of very stable hairpins (this step can be carried out with the PATSCAN software available free at http://www-unix.mcs.anl.gov/compbio/PatScan) (**PATS**). Use the index or search this book for "AT-rich", "GC-rich", "hairpins" to learn what should be done in case these are present (**CDPR**).

I would rather not reproduce the guidelines for primer selection which can be found in every manual on PCR (though different manuals offer somewhat different assortments of recommendations). A few additional tips will possibly let you evade losses of time and labor in the reaction optimization (**BLANK**, **LADD**, **SMEA**, **SPNO**) (**PCDS**).

> **PRIMERS MAY NOT BE GOOD IN PRACTICE EVEN IF THEY ARE GOOD IN DESIGN**

Order another primer pair or try the nested PCR (**NQWP**). Try **PRDI, SBRD**.

> **FOR PRIMER SELECTION USE ONLY DEDICATED SOFTWARE**

Use only dedicated and highly regarded software, such as *Primer3* available free on the internet at http://www-genome.wi.mit.edu/cgi-bin/primer/primer3_www.cgi or *Oligo* (Molecular Biology Insights, USA).

## ➢ DIFFERENCE IN $T_M$ BALANCED BY DIFFERENT PRIMER CONCENTRATIONS

If the primer search is restricted to a short stretch of DNA (e.g., when a primer must be positioned within a short intron), the software could be incapable to find good primers within the acceptable difference in $T_m$ between the primers in a primer pair. Usually, the difference in $T_m$ between the two primers added to the PCR mix to equal concentrations should not exceed 5°C (O. Henegariu, http://www.info.med.yale.edu/genetics/ward/tavi/p02.html). The capacity of the software to find good primers could be enhanced by selecting the primers that have a greater difference in $T_m$ if added to equal concentrations. Subsequently their $T_m$ could be balanced by adding them to the mix to unequal concentrations. In fact, the ten-fold difference in the primer concentration within the recommended range of 0.1-1µM (P. Frame, http://www.ufbi.ufl.edu/~rowland/protocols/pcr.htm) will result only in about 3°C difference in their $T_m$ (http://www.basic.nwu.edu/biotools/oligocalc.html). The additional 3°C difference could sometimes be critical to supply the software with the necessary elbowroom to solve the problem of designing good primers (***BLANK, LADD, SMEA***).

## ➢ INEFFICIENT PRIMING

There is no mispriming, but the primers poorly bind to their proper annealing sites. This happens when the primers are long (exceeding 30 nucleotides) or prone to form primer-dimers or hairpins (***BLANK***). Order shorter and good primers. With long primers try to extend the annealing step (***PPBA***).

## 5. INADEQUATE $MgCl_2$

(***BLANK***, perhaps ***LADD*** or ***SMEA***). See P. Frame, http://www.ufbi.ufl.edu/~rowland/protocols/pcr.htm.

## 6. POOR BUFFER

(*BLANK*). The buffer, i.e., buffer proper and salt plus certain substances (nonionic detergents, gelatin, BSA, dithiotreitol) added to protect Taq polymerase, is adapted to a particular preparation of the enzyme and usually supplied along with the enzyme in a ready-to-use form.

## 7. POOR TAQ

(*BLANK*, *LADD*, *SMEAR*) (**TCBP**).

## 8. THE PRESENCE OF THE PCR INHIBITORS

The natural inhibitor is pyrophosphate building up in late PCR or present in the dNTP stocks under inappropriate conditions of storage (read all of the **KKIM**, **IPMC**), in old frozen solutions, and, possibly, under more or less prolonged exposure to high temperatures (*BLANK*, *LADD*) (**IUYU**). Water. See **AQWW** (*BLANK*). Nucleases. See **ADNA** and **HTHT**. Wood (*BLANK*). Pottering about in the PCR tube with a wooden implement such as a birchen toothpick would inhibit the PCR (**BTIP**). Template (**TEIN**). See **DILU**. Blood, fibrin (*BLANK*, *LADD*, *SMEA*). DNA isolated from coagulated or partially coagulated blood by guanidine-diatoms is, in my experience, a poor template for PCR. Commercially available tubes for blood collection often contain a tiny drop of liquid EDTA in their bottoms. Store them away from heating devices and be sure that the drop has not dried up. Make sure that the nurse who takes the blood immediately and thoroughly (preferably by inversion of the tube) mix it with the anticoagulant. **Blood, heme** (*BLANK*). Heme within erythrocytes inhibits PCR. There are at least three ways to get free of erythrocytes: blood gradient centrifugation, selective lysis of erythrocytes, isolation of the leukocyte nuclei. The reader should estimate their efficiency in removal the pinkish hue from the material prepared for DNA isolation.

**Faeces (*BLANK*).** These contain more than one potent PCR inhibitor [V.S. de Paula et al., 2003, Braz J Infect Dis, 7:135-141]. **Sodium dodecylsulfate (SDS) (*BLANK*).** Naturally, a tiny amount of SDS is often left over in the DNA prepared for PCR. The $10^{-4}$ weight/volume concentration of that substance (0.01%) is sufficient to inhibit the Taq polymerase activity by 90% [R. Cruickshank, http://www.staff.uni-mainz.de/lieb/additiva.html] **(SDSI).** The number of the PCR inhibitors is great. There appears a demand for a catalogue of PCR inhibitors.

## *9. SUBSTANCES THAT DO NOT INHIBIT PCR*

There appears also to be a demand for a catalogue of the substances that do not inhibit PCR and can be present in a PCR mix under any conceivable circumstances (no *BLANK*).

- Agarose and ethidium bromide present therein.
- Some tracking dyes used in electrophoresis can be added to the reaction tube prior to PCR, among them 0.2mM cresol red.
- 12% sucrose and 5% glycerol used to increase the density of the PCR mix in order to enable its transfer into the gel well by a single pipetting.
- Diatomaceous earth, also called "diatoms".
- Cosolvents used in PCR (use the index or search for "cosolvents").

# INADEQUATE STORAGE OF INGREDIENTS FOR THE PCR REACTION

## 1. THE TREACHEROUS REFRIGERATORS

(*BLANK*). The freeze-thaw refrigerators are very convenient. Different models of the freeze-thaw freezers are designed on several distinct principles. Some imply maintaining constant temperature inside the freezer, others allow for a periodic temperature rise above the Celsius zero value. That temperature rise, brief as it may be, would cause thawing of the solutions stored in small volumes and the partial thawing of the solutions kept in larger volumes. The repeated freezing-thawing can be destructive to some chemicals, especially for nucleotide triphosphates. So, when buying a freezer, consult an informed expert and make the correct choice (**IPMC**).

## 2. TEMPLATES, DNTPS, PRIMERS

(*BLANK*) (**TDPB**). The template DNA intended for long-term storage must be thoroughly purified. I have kept my templates, dNTPs, and primers dissolved in water and hardly ever had any problem. Probably, it was just luck, because, on general principles, it is the wrong practice. Deionized water in plastic vessels is acidic. In addition, primers and the extracted DNA can be acids themselves, although purchased dNTP's are usually provided as salts. The acidic pH favors cleavage of the N-glycosylic bond between a purine (adenine, guanine, inosine) and the sugar in a nucleotide. The apurinic site poses an unsurmountable obstacle to any DNA polymerase. If the template DNA represents valuable samples which should be stored for years, I would dissolve it in TE, pH8.0, aliquote, seal the tubes and keep them frozen at -20°C or, rather, at much lower temperature. Its concentration must be the highest available. If the DNA is concentrated, first turn it into a neutral salt because large amount of the acidic DNA may shift the pH of TE buffer. Since TE contains a minor amount of EDTA, the volume of the template added to the reaction should not exceed 1/10 of its total volume (the TOTAL means WITHOUT mineral oil). The purpose of the aliquoting is to reduce the number of times the sample is thawed and frozen, since repeated freezing-thawing could be deleterious to a DNA sample.

(*TRIV*) A broad glass filled with water should be positioned continuously on the floor of the compartment assigned for DNA isolation. (Floor, door handle, pipettes and centrifuges are the most common routes of contamination recognized both in microbiology and the PCR). The water should never be changed, only replenished as it dries. The water with all the dust accumulated therein should serve as substance for the negative controls. Each DNA isolation experiment including a batch of samples must be supplemented with two identical samples taken from the venerable glass. Those mock DNA samples must be treated exactly in the same way and at the same time as the regular samples and put for storage in the DNA sample collection linked with the other samples from the same DNA isolation experiment. Any PCR reaction with the templates from a lot of samples isolated in parallel must include the mock templates obtained in the same isolation – in the capacity of the negative controls.

I would store dNTPs (*BLANK*) and the primers (*BLANK*, *SMEA*, *LADD*) in the same way, only I would not add any EDTA and keep them just in 10mM Tris·HCl, pH8.0. To maintain the final pH close to 8.3-8.8, I would add them to the mix in small volumes. Even so, the stock solutions of dNTPs can have a limited storage time (see P. Frame, http://www.ufbi.ufl.edu/~rowland/protocols/pcr.htm) (**KKIM**). One of the major products of deoxynucleotidetriphosphate degradation is pyrophosphate known to be inhibitor of any DNA polymerase including Taq (**PYRO**, see **IUYU**).

## 3. TAQ POLYMERASE

(**BLANK**, **LADD**). The stock is usually stored at -20°C. The storage medium contains cryoprotector, because it should remain liquid. To avoid freezing do not put it to lower temperature. It is commonly believed that the Taq stock solution should not or need not be mixed before adding Taq to the PCR reaction.

## 4. MgCl$_2$

(**BLANK**, **LADD**, **SMEA**, possibly **SPNO**). See P. Frame, http://www.ufbi.ufl.edu/~rowland/protocols/pcr.htm.

## 5. BUFFER

(**BLANK**, **LADD**, **SMEA**, possibly **SPNO**). The PCR buffer stock is kept frozen at -20°C in small aliquots.

In case you use a ready-to-use mix or "supermix" follow the manufacturer instructions.

# THE THERMOCYCLER

## 1. CONDUCTIVITY OF HEAT PUTS A LIMIT TO THE MIX COMPOSITION

Gross changes in the PCR mix composition may put in disorder the temperature regulation in the thermocycler (***BLANK***, ***LADD***, ***SMEA***).

<u>Note</u>. The positive manner of the next paragraph should not mislead the reader to assume that the author is an expert in thermophysics. The author just believes that if one does not understand anything, it is wiser to face than ignore it. Thus, although my suggested conclusion can be totally wrong, the question itself is justified and worth raising here. Once seeded in the reader's mind, it can stimulate it to investigate the matter and reach the correct answer. All modern thermocyclers are based on the so called active principle of temperature regulation. The principle implies a brief period of accelerated heating/cooling of the metal tubes housing to a temperature above (in heating) or below (in cooling) the temperature programmed for the step. This event immediately precedes the beginning of the step. The magnitude of the temperature over(under)shoot, its duration, and precise mode of the temperature rise and decline are calculated so as to rapidly push the content of the tube just to the programmed temperature. The time count for the step begins as soon as the temperature <u>inside</u> the tube reaches the programmed value. The ground for the calculations is provided by the data on the conductivity of heat of the tube walls and the conductivity of heat of the standard PCR mix [D. Trofimov, pers. commun.] (**CHSP**). The manufacturers of thermocyclers seem to be the only people who definitely know what is taken to be a standard PCR mix, but one can guess that it is a salt solution of the

≈ 0.06 ionic strength (either 50mM KCl+10mM Tris•HCl or 16.6mM $(NH_4)_2SO_4$+10mM Tris•HCl). It is easy to presume that a significant change in the salt concentration or addition of a substantial amount of a cosolvent (that can be up to 15%) will alter the conductivity of heat of the mix, spoil the game of the temperature regulation and cause a poorly predictable play of temperature within the tube. So, an attempt to use a cosolvent or alter the salt concentration can be compared to driving a car in darkness and could possibly be complicated by extensive, quite empiric, and probably fruitless optimization.

At the same time, *Applied Biosystems* (Foster City, CA, USA) does recommend the use of cosolvents on its web site. With regard to the salt concentration, the effect of its alteration seems to be significant only if the alteration is great. With 12.5% and 25% NaCl the difference in the conductivity of heat is 1.23-fold (Handbook on Chemistry, *Goschimizdat*, Moscow, 1951, v.1, p.844, in Russian). 15% and 30% $CaCl_2$ result in 1.05 difference, 11% and 20% $MgCl_2$ in 1.11 difference, 20% and 40% NaBr in 1.05 difference. Although the same source does not provide data on KCl and $(NH_4)_2SO_4$, it seems probable that the slight effect of the salt concentration on the conductivity of heat is a rule common to all or most salts. Therefore, I feel I can do no better than to assume that the usual concentrations of cosolvents and the moderate changes in the salt concentration do not modify to any significant degree the conductivity of heat of the PCR mix, and to adhere to such assumption throughout this book (**WQWH**).

## 2. THE RAMP

(***LADD***; in case of AT-rich sequences, ***SMEA***). In many thermocyclers the ramp speed can be set by the user. I would always make the choice of using the fastest ramp, since a slow ramp means lingering about the temperature set for a step and, in fact, extension of the step under suboptimal temperature (**FQWR**). In some cases there are special reasons for the ramp to be set at the fastest possible. A slow transition from 94°C to 65°C in amplifying AT-rich target would result in considerable time being spent at several degrees over 65°C, leading to dissociation of a growing DNA chain from the template chain and, as a consequence, to incomplete elongation and a smear in the gel (**RQAT**). Perhaps, I am not right. I do not quite understand why the ramp is made to be regulated by the user. If it has to be regulated then there are some situations where a slow ramp is preferable. Furthermore, in one of its manuals to a cycle sequencing kit *Applied Biosystems* does recommend the ramp not to be made too fast. They do not explain why.

## *3. TIME AND TEMPERATURE*

*Time and temperature for the three components of a standard amplification cycle: annealing, synthesis, denaturation.*
Consult manuals but avoid old editions or citations from early papers because the times indicated there refer to the outdated thermocycler models with a matrix temperature regulation. In fact, the modern machines allow for primer extension in shorter than 15 seconds for PCR products less than 400bp long (see explanation in **CHSP**).

A plain practical guide is available in the manual and the troubleshooting guide written by O. Henegariu at http://info.med.yale.edu/genetics/ward/tavi/PCR.html. I won't repeat everything that can be found there. Enough time must be allowed for annealing and extension (***BLANK***, ***LADD***); at the same time it must be just the necessary minimum because excessive time allowed for primer annealing and elongation gives additional chances for the momentary touching of the 3' ends of the primers to complementary or noncomplementary sequence which sometimes results in primer elongation (*LADD*). The optimal annealing temperature is unpredictable but typically 5°C below the primer $T_m$. Often enough it is higher than the calculated $T_m$. An annealing temperature below 50°C is discouraged. At the same time, at extremely low DNA concentrations a drop in the annealing temperature down to 43°C has resulted in enhancement of the PCR sensitivity with no nonspecifics [O. Henegariu, http://info.med.yale.edu/genetics/ward/tavi/p09.html]; with AT-rich template DNA, the optimal annealing temperature can be as low as 37°C [Chevet et al.,1995; *Nucleic Acids Res.*, 23:3343-3344] (*BLANK*). The original DNA rather than the PCR fragments serves as a template in the second and a few subsequent cycles of amplification. For this reason, a high denaturation temperature (about 94°C) must be maintained during the first 10 cycles of a run (I believe, it may also require a longer denaturation time) [E. Rybicki, http://www.mcb.uct.ac.za/pcroptim.htm]. **For shorter fragments**, a reduction of the denaturation temperature/time has been recommended thereafter [E. Rybicki, http://www.mcb.uct.ac.za/pcroptim.htm]. This good advice aiming to save Taq polymerase is almost universally ignored. Prolonged exposure of this enzyme to the very high temperatures usually employed in PCR inactivates it (the published data on the Taq polymerase half-life at 95°C greatly vary [E. Rybicki, http://www.mcb.uct.ac.za/manual/pcroptim.htm#Denaturing, Anonymous, http://arbl.cvmbs.colostate.edu/hbooks/genetics/biotech/enzymes/hotpolys.html]). A deficit of the active Taq can be compensated for by gradual or stepwise extension of the DNA synthesis step in the latter half of the run, or in a few final cycles, or in the last cycle only (*BLANK*) (**DATP**). Consider **ITFM, LGCR, PQFD, EQWT, TSPCR, DELT, TMAC, RQAT, FQWR, SBCYC, REME, HEDE, SOLT, PTTM, TOUC, IFNO, EVRESP, ATAS, SDVS, PPBA**.

## 4. DUSTY, GREASY, OR FLUFFY TUBE WELLS

(***BLANK***, ***LADD***, ***SMEAR***). As the thermocycler temperature regulation becomes more precise and uniform from well to well, there is evidently a greater demand for a close contact between a tube and a well. Fill the well with ethanol. Rub it with a pipecleaner or interdental toothbrush. Suspend the insoluble dust particles (which can be unseen but present) by pipetting and quickly (before the dust has had the time to settle down again) aspirate the ethanol away from the well. Let the well dry up completely and then inspect it with a magnifying glass for any fluff left behind by the tools you used. The dust has a habit of falling downwards in a continuous fashion, so make it *your* habit to keep the well block covered. (NEVER add any lubricant to a well of a modern thermocycler, that is, a machine which asks you a question about the volume of the mix before each run) (**DGFTW**).

## 5. RAPID EVAPORATION

Rapid evaporation of the tiny volume of liquid not overlaid with oil, above the paraffin or directly from the tube. Inserting the tubes into the hot thermocycler with heated cover or taking them out of it is not as brief as it may seem (***BLANK***) (**EVAP**).

## 6. SUBOPTIMAL PERFORMANCE OF THE THERMOCYCLER OR ITS PARTICULAR WELLS

(***BLANK, LADD, SMEA, SPNO***) (**SPTPW**). Run the self-check functions of the machine. Once or twice a year summon an engineer to thoroughly examine the apparatus. Have a couple of thermocyclers obtained from two different vendors, each having at least ten years experience in the development and production of this sophisticated and precise machinery. As a precaution against poor well-to-well reproducibility use the same wells between different experiments. To be on the safe side, avoid peripheral rows of wells.

## 7. THE TUBES ARE POORLY PRESSED DOWN INTO THE WELLS OR THEY HAVE GOT DEFORMED

(*BLANK, LADD, SMEA, SPNO, HMWG*) (**TKKT**).

# FAULTY TARGET SELECTION

(*BLANK, SPNO*) (FTSE)
The PCR is typically designed to be applied to any of the related genomes (representatives of a species, genus, kingdom, etc.). It must be specific to the **selected set of genomes**, that is, the genomes outside the set but expected to be present in the examined samples must not contain the target for the amplification. The *target* can be operationally defined as a nucleotide string of fixed length flanked by primer annealing sites which must be identical between the genomes of the selected set but different to sequences in the genomes outside the set. The sequence in the interprimer region may vary – as far as the constant length is maintained, if the method for detection of the PCR product is electrophoresis.

An initial step in the target selection can be a computer search for the sequences in the genomes within the set that are similar enough to contain areas of complete identity at the intervals corresponding to the length of the PCR fragments. This could be done with the BLAST multiple alignment program which screens the immense number of nucleotide sequences located in the Genbank database [http://www.ncbi.nlm.nih.gov]. The BLAST outcome will also contain answers to the crucial question of whether the suitable regions of complete similarity are shared with the genomes outside the set expected to be present in the samples. The BLAST result will be as close to the reality as the Genbank resources are close to it. As soon as the specificity of the identical regions is verified, primer selection can be carried out. A specimen sequence should be imported into a primer selection software and the computer should be instructed to search for primers within the appropriate regions of the complete similarity.

It should be noted that there may be additional, and undesirable, primer annealing sites within the same genome. The primer selection software usually discards the primers which have at least partial homology to the repeats found in several eukaryotes (usually man and mice). You can check the result and extend it beyond the software capacity by using BLAST against the Alu database containing repeats from many organisms.

Finally, the ends of the Genbank entries can contain vector sequences. So, discard a dozen of nucleotides from each end and check the designed PCR fragment against another database called "Vector". See **CDPR** and **PCDS**.

Use of degenerate primers is out of the scope of this book [see E. Rybicky, http://www.mcb.uct.ac.za/manual/pcroptim.htm].

*Michael L. Altshuler*

# INCOMPLETE DNA DENATURATION AND DISPERSAL

## 1. TEMPLATE DNA

In some cases, the template DNA can be very resistant to denaturation (***BLANK*** or ***LADD***). If the 5 minutes at 95°C or 15-20 seconds at 97°C in the beginning of the process does not help (**ITFM**), try the simplest alternative – reduce the template concentration to a minimum (if that was initially high) (**MINI**). Yet, it may still not be effective. Denature the template DNA in TE buffer, pH8.0. TE is well suited for DNA denaturation because of its low salt concentration. Since the purified DNA is usually dissolved in TE, the only preparatory operation you may have to do is to dilute it as much as feasible. Diluted DNA is easier to denature than the concentrated DNA. Heat the DNA solution to 90-95°C for 5 minutes and then flash-cool it by an instantaneous transfer of the tubes into snow or ground ice. Repeat it two or three times, and keep the tubes tightly closed to prevent evaporation and contamination. Note that the temperature shown on the display of the heating device can be quite different to the temperature inside the tubes, if the tubes do not closely fit into the wells of the heated metal block. (In fact, every point on the surface of the tube down the well must touch the well wall. The most reliable tube heater is a thermocycler.) You can further enhance denaturation by adding cosolvents of the helix-destabilizer class (see **HEDE**) both to the DNA solution and to the PCR mix. To destabilize the double helix you can also reduce the salt concentration in the PCR buffer or the concentration of the whole of the buffer (see the references under the code **SKCI**) (**SQWM**).

To exclude the possibility of the DNA clustering or clotting, to ensure rapid diffusion of the ingredients throughout the reaction volume and, again, to facilitate DNA denaturation, the DNA must be fragmented. The level of the original DNA fragmentation depends on the method of DNA isolation. Isolate DNA with diatoms rather than with phenol/chloroform because the somewhat harsh conditions of the former method promote breaking very long DNA strands [D. Trofimov, pers. commun.] (**MDIS**). A way to tear the DNA strands is sonication. Sonicate them in TE, pH8.3, under a variety of conditions and check the results by electrophoresis (see **KLMN**). Alternatively, fragment the template with a restrictase which has no sites within the target, desalt the digest and then dissolve it in TE, pH8.3-8.8. Heat and flash-cool it several times. The shorter the DNA fragment is, the easier it denatures, and the easier it renatures. The renaturation is accelerated when the DNA concentration is great, and the renaturation progresses with time. So put the tubes from ice into a freezer or start your PCR without much delay (**RQTE**).

Whether it is **ITFM**, **MINI**, **SQWM**, **MDIS** or **RQTE**, increase denaturation time during the first few cycles, irrespective of the length of the expected PCR fragment (this point is elaborated under **DATP**).

## 2. PCR FRAGMENTS

(***BLANK*** or ***LADD***). The longer (up to 4-5kb) or GC-rich PCR fragments require longer denaturation times or higher denaturation temperatures (perhaps up to 1min at 94°C if the cycle number is the usual 35, and even longer, if the number of cycles is any less) (**LGCR**). Too long denaturation times with 35 cycles would result in Taq inactivation and premature termination of the PCR process (see **DATP**). Helix destabilizers (see **HEDE**) could allow the complete denaturation during shorter time, saving Taq and thereby increasing the sensitivity and the efficiency of the reaction (**PQFD**). See **SKCI**.

Denaturation of the PCR fragment is opposed by its renaturation at the annealing or synthesis steps but the efficiency of the renaturation is negligible as compared to the efficiency of the primer annealing until the concentration of the PCR product is very close to its maximum [E. Rybicki, http://www.mcb.uct.ac.za/manual/MolBiolManual.htm]. The renaturation puts a stop to the accumulation of the fragment (**CQBA**).

## 3. HAIRPINS

(***BLANK*** or ***LADD***) (**HAIRP**). *Hairpins in the internal (inter-primer) region of a PCR fragment.*
Dissociated DNA strands will not reassociate at the annealing temperature unless the concentration of the PCR-fragment is very high. But the stable hairpins will. The presence of hairpins in the template strand at 72°C will obstruct DNA synthesis. Thus, the hairpins capable of reconstruction at the annealing step present a particularly difficult problem. See **PATS**.

➤ It can be hopefully solved by the addition of efficient helix destabilizers (see **HEDE**).

➤ Increasing the elongation time (**EQWT**).

➤ Cautious reduction of the salt concentration in the PCR buffer or reduction of the buffer concentration as a whole (see **SKCI** and the references under that code).

➤ The employment of the two-step PCR with long primers (remember that longer primers require somewhat longer annealing times), the temperature of the lower step exceeding 72°C by several degrees (up to 78°C) (**TSPCR**).

➤ Take additional pains to ensure the purity of the mix (especially the DNA) and the integrity of the triphosphates.

# THE TAQ ENZYME

## 1. HURDLES FOR TAQ POLYMERASE

➢ **STABLE HAIRPINS IN THE TEMPLATE STRAND**
(*BLANK* or *LADD*). See **HAIRP**.

➢ **AT-RICH AREAS**
(*BLANK* or *SMEA*, less probably *LADD*). See **PATS**. Within the AT-rich runs, if the elongation phase is carried out at the usual 72°C, the nascent DNA strand dissociates from the template strand and its growth stops. Thus, the AT-rich PCR fragments often appear as a barely visible band with a smear beneath it (**HEST**). Decrease the elongation temperature (it can be something like 65°C or lower) (**DELT**). Try tetramethylammonium chloride (see **TMAC**). Mind **FQWR** and **RQAT**. To stabilize the double helix increase the salt concentration in the buffer (see **SKCI** and the references therein).

➢ **GC-RICH AREAS**
(*BLANK* or *LADD*). See **PATS, LGCR, DATP, HEDE, PQFD**. Make especially rigid denaturation in **MINI, SQWM, RQTE, PQFD**. The saturation in the concentration of the PCR fragment due to renaturation of its strands occurs at its lower concentration (see **CQBA**). To destabilize the double helix decrease the salt concentration in the buffer (see **SKCI** and the references therein).

➢ **ALTERNATING GC/AT-RICH REGIONS**
(*BLANK, LADD, SMEAR*). See **PATS**. Try subcycling [Liu et al., 1998, *Biotechniques*, 25:1022-1028] (**SBCYC**).

## 2. HOT START

(**BLANK**, **LADD**, **SMEAR**). The hot start implies suppression of the Taq polymerase function at room temperature during the assembly of the PCR mix and through the initial heating of the reaction mix up to the annealing temperature. Hot start is a powerful means to increase the reaction specificity, sensitivity and the yield. The original hot start method required the addition of Taq to preheated tubes separately to each one, without taking them away from the thermocycler. It was inconvenient, and the concentration of the enzyme was likely to be less even between the tubes than it would have been if it had been added to the mastermix. In addition, with the solutions being hot, that operation increased the risk of contaminating the environment with the template DNA. This brought forth a demand for improved and closed-tube hot start.

### ➢ THE IMPROVED HOT START

Some companies market Taq polymerase sealed in wax granules which melt during the initial heating (*AmpliWax PCR Gems,* Applied Biosystems, USA). The common in-house contraption involves a plug providing a barrier between Taq polymerase and other components of the mix. The plug is usually made of the paraffin-mineral oil mix which is adjusted to a melting point of about 75°C. Another trick is to prepare a complex between the Taq protein and the anti-Taq antibody. As the temperature goes up the antibody denatures and releases the active Taq molecules (*JumpStart Taq DNA Polymerase,* Sigma-Aldrich, USA). A similar approach might have been employed by Roche Applied Science (*FastStart Taq DNA Polymerase*) and Finnzymes Oy (*DyNAenzyme II Hot Start DNA Polymerase*). Instead of manipulating the enzyme the USB company has found a protein which sequesters primers (*HotStart-IT Taq DNA Polymerase*). The protein is thermally unstable and releases the primers at a sufficiently high temperature.

Any of the aforementioned methods implies that the original Taq capacity to function at low temperatures is restored by the first step of annealing. This is good, if the annealing temperature has to be low (the primer extension **must begin** at the annealing temperature [E. Rybicki, http://www.mcb.uct.ac.za/manual/pcroptim.htm#Annealing]. The Eppendorf company has invented a thermostable Taq polymerase inhibitor which completely blocks the Taq function below 40°C, allows for a steep rise in the activity in parallel with the temperature rise from 40°C to 55°C, and does not interfere with the activity above 55°C (*The Hotmaster Taq DNA Polymerase Kit*). This kind of the temperature dependence is reproduced at every cycle throughout the amplification run. In some cases the Eppendorf innovation may have special benefits. When the concentration of both target and nontarget DNA is reasonably high (see Table 1, page 18), a sharp reduction of the Taq activity at the annealing temperature could diminish the formation of the nonspecific products without compromising the yield.

## ➤ THE ROLE OF THE REACTION VOLUME IN THE QUASI HOT START

(***LADD***). There is the original open-tube version of the hot start, expensive commercial closed-tube hot start preparations and the in-house closed-tube hot start (see the preceding subsection). Some practice also a well-known trick of the quasi hot start by putting the tubes containing all ingredients including the enzyme into a preheated thermocycler. (It would be safer to preheat it with closed or screwed up lid and at least one empty tube put into the machine; it would be better to keep the tubes on ice during assembly of the PCR mix and wipe them dry upon removal from the ice). The results obtained with the quasi hot start are generally not as good as the results of the bona fide hot start. The quasi hot start is worth mentioning here because of a subtlety which seems to be seldom realized by anybody.

*The less the volume of the reaction is, the closer the quasi hot start is to the real hot start because the process of the heating is more rapid.* Mind **EVAP**.

## 3. NONSPECIFIC BINDING OF TAQ TO DNA

(***LADD, SMEA***). A drawback of the Taq polymerase is its propensity to bind or coat PCR fragments precluding subsequent manipulations with them (e.g., restriction). Furthermore, the Taq-DNA complex must have altered electrophoretic mobility, a phenomenon known to molecular biologists under the designation of "mobility shift" and employed to identify proteins bound to particular DNA sequences (***PBOI***).

The extent of the Taq binding to PCR fragments varies as assayed by their resistance to restriction. With some Taq preparations the restriction is complete, with others a small portion of the PCR fragment remains resistant, and that portion is always the same at different incubation times and concentrations of the restrictase [M.L. Altshuler, unpublished]. According to published information, there are cases when all DNA molecules are densely covered or coated with the protein [E. Rybicki, http://www.mcb.uct.ac.za/manual/MolBiolManual.htm]. It is difficult to account for such variation, but for one thing: Taq polymerase is the only component of the PCR mix which is added with regard to its activity rather than its quantity. Thus, the amount of the added protein, both active and inactive, may determine the degree of the binding. Yet, the situation remains paradoxical. As a result of a PCR we usually get a DNA band taking a position in the gel, as if there were no protein bound to it. On the other hand, the evidence is abundant that in a good many cases it is, in fact, a piece of DNA tightly bound or coated by the packed Taq protein molecules. If such were the case, its electrophoretic mobility must have been shifted according to the net molecular weight and the combined

DNA-protein charge. Does the protein compete with ethidium along the way for a place on the DNA and is peeled off by the dye?

# INCOMPLETE PRIMER ELONGATION OR PREMATURE TERMINATION OF DNA SYNTHESIS

(**IPEL**). Premature termination of DNA synthesis frequently occurs when the PCR process approaches its termination, although in some cases it can take place earlier. Its evident result is the generation of partially double-stranded molecules. Its less evident outcome could be the formation of the molecular aggregates and networks.

## 1. UNDER-ELONGATION OF PRIMERS IN THE LATE PCR

(***BLANK***, ***SMEA***, possibly ***LADD***). This problem seems to present itself predominantly with the relatively long PCR fragments. Insufficient concentration of the active Taq, primers and dNTP's together with the inhibitory action of pyrophosphate relative to the elevated amount of the accumulated PCR fragment is the generally accepted reason why many DNA strands remain unfinished. Incomplete primer elongation in late PCR is also probable when the formation of primer dimers competes with the generation of the desired product. Reannealing of the PCR product strands could be a factor preventing completion of a DNA chain (see **CQBA**). The unfinished strands may remain as waste or serve in the capacity of primers in the subsequent PCR cycles. In the gel the mobility of the partially double-stranded PCR fragments must be different from the mobility of the fully double-stranded molecules.

There is also a possibility of renaturation of the complementary strands of the PCR product, if they remain partially single-stranded. The renaturation must be favored by high concentration of the DNA and by the slow, delayed or interrupted cooling of the PCR mix after the termination of the PCR process. It is also time-dependent and can slowly proceed at room temperature. Moreover, large clusters of annealed DNA strands may be formed according to the following model. The single-stranded region probably associates with a complementary single-stranded portion of another partially elongated PCR product leading to the formation of complexes containing up to four DNA strands: two complementary DNA strands of full length annealed to each other in the middle of the amplicon, each carrying an underelongated primer on its 5' end. The other end of each full-length DNA strand remains as a single-stranded overhang. The four-stranded structure could serve as a nucleus for the generation of a larger cluster by way of the single stranded overhangs anchoring more molecules. If that is true, it could be easily imagined that under favorable conditions the formation of the molecular aggregates might be very efficient. They would be visible in the agarose or polyacrylamide gel, would have altered electrophoretic mobility and would be misinterpreted as nonspecifics (**LADD**).

Increase the time for DNA synthesis in late PCR, especially in the last cycle (**REME**). If you do not care for the yield, reduce the number of cycles – use SYBR GOLD (*Invitrogen-Molecular Probes*) or any other sensitive method of detection – in the end, the work can turn out to be the cheaper option and, in many instances, the result is closer to the reality both qualitatively and quantitatively.

Increase the concentration of nucleotides + $MgCl_2$, primers, Taq. Check the integrity of nucleotides and the purity of the template DNA. Alternatively, make the DNA chains be completed by opening the tube, adding new ingredients, running a single additional cycle of PCR, preferably with a prolonged elongation step [Delwart et al., 1993, *Science*, 262:1257-1260]. Evidently, the latter advice can be only used if the thermocycler is positioned in the contaminated or "dirty" compartment of the laboratory.

This might be the right place to add that, if the DNA synthesis step in the last cycle is made unusually long or the final cooling is slow, the formation of the multi-strand molecular aggregates is conceptually possible even when the primer elongation is complete and the PCR product is represented by the fully double-stranded molecules. "DNA breathing" or the momentary unwinding of the DNA ends might account for the joining of the molecules into chains. It is not a mere theory [M.L. Altshuler, unpublished]. Try to interrupt the final cooling by 1 min at 55°C and take two aliquots of the PCR product. Denature one aliquot and leave the other as it is. Then run them in nondenaturing polyacrylamide gel and stain the gel with silver. An amazing multitude of fine bands will be seen throughout the nondenatured sample lane; the denatured sample lane will be indistinguishable from the ordinary result of an SSCP experiment. So, to be on the safe side, do not extend the last elongation cycle too far and always employ the forced cooling method after the termination of a run (*LADD, SMEAR*) (**ATAS**).

## 2. PREMATURE TERMINATION OF THE DNA SYNTHESIS

Premature termination of the DNA synthesis is a characteristic feature or optional by-effect of many situations considered elsewhere in this book. See for yourself what can or may happen in **EDNE, APPT, MPRE, KPPT, PRST, READ, PQFD, NQWP, TCBP, IUYU, TEIN, TDPB, CQBA, HAIRP, HEST, SBCYC, WQWH, RQAT, DGFTW, EVAP, SPTPW, TKKT, ADVI, ALDR**, etc. To counteract premature termination take the advice associated with the codes listed in this paragraph or elsewhere in the book.

# COSOLVENTS OR PCR ADDITIVES OR PCR ENHANCERS

Cosolvents are chemicals added to the PCR mix. They can have a beneficial effect on the outcome of amplification with regard to its specificity and sensitivity. Whether they **will** have it, is unpredictable (R. Cruickshank, http://www.staff.uni-mainz.de/lieb/additiva.html). The mechanisms of their action are often unknown. Many researchers erroneously identify all cosolvents with the agents that facilitate DNA denaturation (helix destabilizers). In fact, the mechanism of their action is broader. Conceivably, cosolvents may be especially helpful with longer PCR fragments. Different cosolvents can be *combined*. Mind the pH of a cosolvent and the final pH of the reaction mix (***BLANK***, ***LADD***, ***SMEA***).

## *1. HELIX-DESTABILIZERS*

(**HEDE**). Helix destabilizers are the substances which promote DNA denaturation, i.e., dissociation of DNA strands in a DNA double helix. The same substances evidently must and do prevent the annealing of primers to their targets reducing the priming efficiency and diminishing the yield or sensitivity of PCR. Nevertheless, a mere addition of a helix destabilizer without concominant compensation for the inefficient priming could result in the enhancement rather than reduction of the reaction yield as well as in the enhancement of the specificity (***BLANK***, ***LADD***, ***SMEA***). How to compensate for the reduced priming? Decreasing the annealing temperature appears to be good advice in cases when helix destabilisers are required to facilitate DNA denaturation or prevent its renaturation at temperatures higher than the annealing temperature. In those cases when it is essential to destabilize the helix at low temperature or prevent the formation of heteroduplexes at the final cycles of the PCR try to raise the primer concentration or use longer primers concomitantly with the addition of a helix-destabilizer with or without increasing the annealing temperature (***BLANK, LADD, SMEA, SPNO***) (see additional advice under **HAIRP** or **HETE**). (**SOLT**). Examples of the PCR-compatible helix destabilizers include **pure** formamide, **unoxidized** dimethyl sulphoxide (DMSO), dimethyl formamide, urea, glycerol, 7-deaza-dGTP, betaine (mono)hydrate and various proprietary merchandise. Chakrabarti and Schutt have udertaken a search for new helix destabilizers and have succeeded in the discovery of several compounds which outperform those just mentioned [Chakrabarti and Schutt, 2001, *Nucleic Acids Res.*, 29:2377-2381; Chakrabarti and Schutt, 2002, *Biotechniques,* 32:866-874; Chakrabarti and Schutt, 2001, *Gene,* 274:293-298]. Tetramethylammonium chloride has been found to be a "miracle" substance enhancing both sensitivity and specificity of amplification of various targets with ***any*** GC-content [Chevet et al.,1995, *Nucleic Acids Res.*, 23:3343-3344; Hung et al., 1990, *Nucleic Acids Res.*, 18:4953]. In DNA with normal GC content I have failed to find examples confirming that. In my hands it behaved in a manner typical to a usual helix destabilizer, i.e., I had to reduce the annealing temperature to restore the yield obtained in the absence of tetramethylammonium chloride.

## *2. HELIX STABILIZERS*

(***BLANK***, ***LADD***, ***SMEA***) (**TMAC**). It is hardly possible to amplify AT-rich DNA without the aid of an agent capable of strengthening the double helix. Tetramethylammonium chloride does it. It allowed amplification of an 80% AT DNA fragment enhancing both specificity and sensitivity [Chevet et al.,1995; *Nucleic Acids Res.*, 23:3343-3344]. What was its precise role? To allow the correct primer annealing or to make feasible the primer elongation (see **HEST**)?

## 3. SUBSTANCES THAT NEUTRALISE THE PCR INHIBITORS

(*BLANK*). See **SDSI**. 0.5% NP-40 or Tween-20 will counteract the PCR inhibition caused by SDS. Bovine serum albumin can prevent the inhibition by melanin, a substance present in the melanoma cells (http://www.staff.uni-mainz.de/lieb/additiva.html).

## 4. PCR ENHANCERS WITH POORLY UNDERSTOOD MECHANISM OF ACTION

(*BLANK*, *LADD*, *SMEA*). These include nonionic detergents added at the concentrations exceeding those usually present in the PCR buffers [R. Cruickshank, http://www.staff.uni-mainz.de/lieb/additiva.html]. Similarly, bovine serum albumin at unusually high concentrations has been found to be an efficient enhancer of the PCR yield [O. Henegariu, http://info.med.yale.edu/genetics/ward/tavi/p16.html].

For further reading on PCR enhancers the reader is referred to http://www.pcrlinks.com/generalities/additives.htm

# APPROACHING THE LIMIT OF THE PCR SENSITIVITY

### (*BLANK, LADD, SMEAR*) (ALPS)

In certain cases PCR has to be very sensitive; in others there is no need for high sensitivity. Nevertheless, almost every paragraph in this book in fact addresses the issue of sensitivity.

The difficulty of getting highly sensitive PCR increases in proportion to the ratio of the nontarget to the target DNA. Consequently, the utmost difficulty is to amplify 1 copy of the target sequence to a microgram level in the pesence of about 1 µg of nontarget DNA. For a 500bp PCR target this corresponds to a multiplication factor of $10^{12}$, i.e., a conversion of 0.5 attogram to 0.5 microgram (see Table 1, page 18, and the notes following it). This section brings together the points believed to be especially important with regard to the enhancement in sensitivity.

1. It is generally accepted that the crucial point of PCR is the few initial cycles of amplification. Very low copy number of the target would make the encounter of the primers with their annealing sites a rather rare event. The same consideration applies to the encounter of the Taq polymerase molecules with the 3' ends of the annealed primers or the 3' ends of the nascent DNA chains (several Taq molecules have to be exchanged for completion of a synthesis of a single DNA chain since this DNA polymerase has a limited processivity). To allow enough time for those encounters to occur, it is advisable to prolong the primer annealing step in the first cycle (or, perhaps, in a few initial cycles) of a run up to 3 minutes and the DNA synthesis step up to 5 minutes [E. Rybicki, http://www.mcb.uct.ac.za/manual/pcrcond.htm] (probably, with modern thermocyclers provided with active

thermoregulation this time can be reduced, see **CHSP**). The prolongation is done at the evident risk of exchanging *BLANK* for *LADD* because the more time that is allowed for annealing and extension, the more opportunity is provided for nonspecific priming. So a higher annealing temperature during the prolonged annealing (in fact, the touchdown) has been recommended by the same authors (**PTTM**).

Slow diffusion of ingredients owing to elevated viscosity of the mix would aggravate the encounter problem. The elevated viscosity can be due, for example, to the large general amount of DNA or the presence of glycerol in the mix. In the former case, DNA fragmentation can be helpful (see **RQTE**, **MDIS**).

Raising primer concentrations to a higher level may be deemed an alternative to the step prolongation; raising Taq concentration can be also attempted.

2. Employ the hot start.
3. Employ the touchdown, making the final annealing temperature as low as it is possible without getting *LADD* (**TOUC**).
4. Increase the number of cycles, at the same time taking care to save Taq polymerase (see **DATP**).
5. Shorter primers have greater priming efficiency [Anonymous, General Notes on Primer Design in PCR http://www.biocenter.helsinki.fi/bi/Programs/oligoinfo.htm].
6. If possible, make the length of the PCR fragment something like 500-800bp. Short fragments will require high molar concentration to be visible in the gel and, therefore, higher multiplication factor; longer fragments will require more optimization.
7. At extremely low total DNA concentrations a sharp drop in the annealing temperature may enhance the sensitivity without getting nonspecific bands or smear [O. Henegariu, http://info.med.yale.edu/genetics/ward/tavi/p09.html].

8. Reduce the volume of the reaction mix down to 5μl [O. Henegariu, http://info.med.yale.edu/genetics/ward/tavi/p03.html].
9. Do nested PCR or reamplification.
10. According to O. Henegariu [http://info.med.yale.edu/genetics/ward/tavi/p09.html], at the same nontarget/target DNA ratio the small **total** amount of DNA would aggravate the problem of specific amplification. So, increase the amount of both target and nontarget DNA by amplification with random primer [for references search PUBMED at http://www.ncbi.nlm.nih.gov/entrez/query.fcgi?db=PubMed with the keyword "whole genome amplification"].

Selective amplification of cognate sequences, e.g., mutated *ras* genes from a sample containing a great number of normal cells and a few cancer cells, is beyond the scope of this review and pertains to the theme of detecting mutations and polymorphisms with PCR or in PCR products.

# AGAROSE GEL ELECTROPHORESIS

## 1. THE BAND DIFFUSES AND DISAPPEARS

(*BLANK*) The band diffuses and disappears because electrophoresis is too slow or too long, especially when the fragment is too short for the given gel concentration and the gel is thin. Pour the gels as thick as possible – this keeps the DNA enclosed inside the gel and provides the additional benefit of preventing the DNA from entering the environment and causing contamination (D. Trofimov, pers. comm.)

## 2. SHORT FRAGMENTS OF UNEVEN LENGTH MIGRATE DOWN THE GEL WITHOUT ANY SEPARATION

(*PBOI*) Too low gel concentration.

## 3. THE BAND IS INVISIBLE

(*BLANK*) The band is invisible because EthBr has gone out from the bottom part of the gel. Add it both to the gel and to the buffer.

## 4. THE SIGNIFICANCE AND THE INSIGNIFICANCE OF THE SALT CONCENTRATION IN THE COMPARED SAMPLES

(*SMEA*, *HMWG*, *PBOI*). Suppose you load into the gel two samples containing identical DNA fragments. There is high salt concentration in one sample and low salt concentration in the other. At first (usually during 5 to 15 minutes) the fragments from the high salt sample will lag behind the fragments from the low salt sample. Then the high salt fragments would speed up and finally catch up with the fragments from the low salt sample. Thus, they will take the relative position they must have occupied from the beginning, had the salt concentration been the same in both samples. A situation of this kind is often observed when the salt concentration in the PCR buffer differs from the salt concentration in the DNA molecular weight standards. So, do not despair but wait a little (**HREN**).

## 5. THE BANDS SMEAR DUE TO THEIR FAST MOVEMENT OR THE DNA OVERLOAD

(*SMEA*) Electrophoresis is too fast – bands smear. Excessive amount of DNA loaded in a gel well will also result in smearing.

## 6. DIRTY GEL SUPPORT

(*UNTIDY*) Dirty gel support. Wash and wipe it.

## 7. DRIED WELL

(*HMWG*, *GHOS*) As soon as you remove the comb put the gel in buffer and, in general, do everything to prevent the drying-up of the wells. They must remain wet.

## 8. DNA STICKING IN THE WELL CAUSED BY INAPPROPRIATE GEL DENSITY

(*HMWG*) The molecular weight of the loaded DNA is too high for a given gel concentration.

# CAUSES FOR SPECIFIC NONSPECIFICS AND THE FALSE CONTAMINATION

## 1. CHIMERAS

(**SPNO**). The generation of chimeric or recombinant PCR products can occur to a significant degree in diploid cells or in samples containing DNA from a mixture of cancer and normal cells. Incomplete elongation of primers is assumed to lie at the hub of this PCR artifact. The under-elongated primer carrying a portion of one species of the target is annealed to another species in the subsequent cycle. Its further extension leads to production of the recombinant DNA strand. A clear explanation of how this comes about is presented by Judo et al. [1998, *Nucleic Acids Res.*, 26:1819]. Every precaution should be taken to avoid incomplete primer elongation (search for **IPEL**). In particular, the number of the cycles should be reduced using, perhaps, real-time PCR or a very sensitive dye such as SYBR Gold (*Invitrogen-Molecular Probes*). Provide enough time for the elongation to be completed increasing the elongation time close to the end of the process. Use as short PCR fragment as possible. Provide enough good quality dNTP (400mM each for a 590bp PCR fragment in the work of Judo et al.), enough primers, Taq, and ensure the purity of the reaction mix (**ADVI**).

## 2. ALLELE DROPOUT

(**SPNO**). PCR carried out on DNA isolated from diploid cells must result in 1:1 mixture of the two alleles. However, owing to a variety of causes, one allele is often amplified faster than the other resulting in the lack of one of the alleles in the outcome of the PCR. Reading this book you will quickly spot a range of situations which can occasion the allele dropout. Examples include a difference in allele length, unequal denaturation, hairpins in one allele, AT-rich runs in one of the alleles, etc (**ALDR**).

## 3. *HETERODUPLEXES* (HETE)

(***SPNO***). Heteroduplex is a double-stranded DNA molecule, the strands being similar but not identical. When the target for the DNA amplification is represented by a set of similar but nonidentical molecules, such as alleles in heterozygous diploid cells, the result of the PCR would include heteroduplexes. Heteroduplexes result from the strand reannealing at or after PCR. At the final cycles of amplification the concentration of the PCR product becomes so high as to favor reannealing of the complementary and partially complementary strands competing with the primer annealing. The strand reannealing is aggravated by conditions causing incomplete or slow elongation of primers (see **IPEL**). If the PCR is done in order to detect a DNA fragment of a defined length with a given pair of primers or to prepare material for the diploid cell sequencing, the formation of heteroduplexes can be ignored. In other cases they are not wanted. To avoid them, reduction in the number of the PCR cycles would be a good choice, though it will compromise the yield. To counteract incomplete elongation of primers see the advice under **IPEL**. A radical way to get rid of heteroduplexes is to "resolve" them [R. Delwart et al.,1993; *Science*, 262:1257-1261] by diluting PCR product in the PCR buffer, adding fresh dNTPs, primers and the enzyme, and then running one or a few additional cycles with prolonged elongation time. Extension of the DNA synthesis step in the last cycle followed by rapid final cooling down to 4°C could be helpful against heteroduplexes (**IFNO**).

Regarding the electrophoretic mobility of heteroduplexes, it will be the same as that of the homoduplex if the length of the strands is precisely the same. Insertion or deletion of just one nucleotide in one of the strands would cause the DNA molecule to bend. It may affect the speed of its movement along the agarose gel.

## 4. PRIMER MULTIMERS

(*TRIV*). Do not entirely forget that contamination (the presence of the PCR fragment in the negative control) may be false and be a result of a primer multimer of the appropriate size. Such cases may not be as infrequent as they would seem. It appears natural that the false product would be observed in the negative control only, since in the absence of the specific target the balance of the PCR reaction is shifted towards the generation of any kind of the products resulting from primers. Reduce the primer concentration. Try **EVRESP** or **NQWP**.

## 5. LOW RESOLUTION ELECTROPHORESIS RESULTING IN IMPRECISE IDEA OF THE CORRECT BAND POSITION

(*SPNO*). The distance between the bands in the 100- or 1000bp ladder in the marker lane must be great.

## 6. COINCIDENCE OR THE DEVIL

(*SPNO*). Find the devil out by nested PCR, restriction analysis or by sequencing.

## MINERAL OIL AND WAX
(*BLANK* or *SMEAR*)

### 1. MINERAL OIL

Avoid exposure of mineral oil to UV. Avoid autoclaving mineral oil. Use high quality mineral oil which is not contaminated with nucleases. See P. Frame, http://www.ufbi.ufl.edu/~rowland/protocols/pcr.htm.

### 2. WAX, PARAFFIN OR VASELINE

Remember that wax, paraffin or vaseline can contain ions, nucleases and other PCR inhibitors (*BLANK*, *SMEAR*).

# PRIMER-DIMERS AND PRIMER MULTIMERS

(*BLANK*, *LADD* or *SPNO*) (**PRDI**)
There seems to be no universally adopted definition of the term *primer-dimer*. Some consider any stable association between two primer molecules to be a primer-dimer, while others are convinced that the true primer-dimers result from the action of the DNA polymerase upon the associated primers. Fig. 1 and 2 on the following pages exemplify the diversity for possible definitions and conceptions of this term. "Primer-dimers" are observed in every other amplification. Sometimes, primers also join together into long DNA fragments representing repeats of a primer sequence [LF Wang et al., 1997; "Head-to-tail primer tandem repeats", Mol. Cell Probes, 5:385]. How it comes about, is unknown.

Despite the fact that the partially complementary interprimer aggregates are less stable than the fully complementary primer-target complexes, the high primer molar concentration will rise their $T_m$ and increase the likelihood of their generation in the range of the annealing temperatures.

The formation of the primer aggregates and hairpins of any kind is undesirable. Those that can serve as substrate for the DNA polymerase are especially unwanted because every elongation event will irreversibly ruin one primer and enhance the complementarity between the extended and the original primer. Those that cannot serve as substrate for DNA polymerase will diminish the number of the primer molecules available for annealing to the PCR target. See **GVPT**.

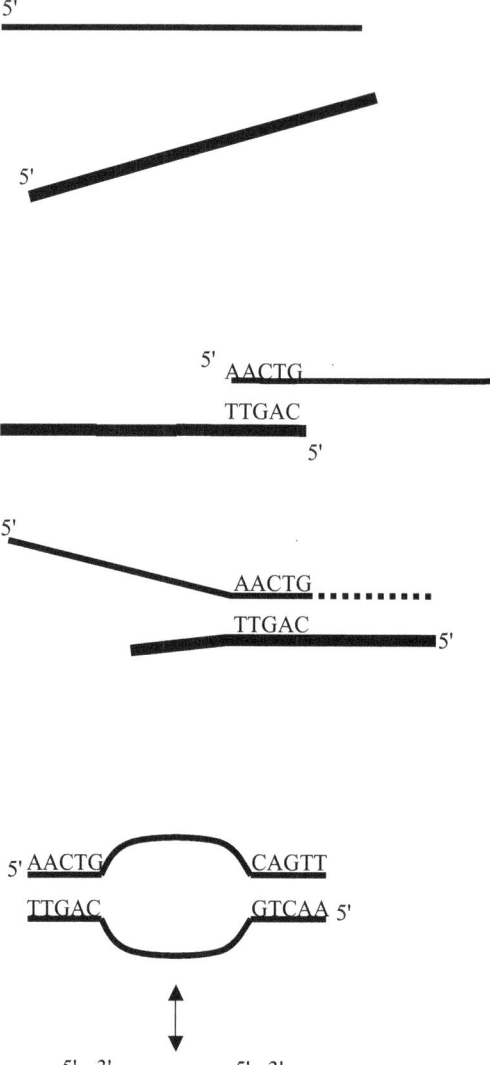

**Figure 1. Some examples of the primer-primer associations and the hairpin structures.**

A thin solid line stands for a sequence of nucleotides in one primer. A thick solid line stands for a sequence of nucleotides in the other primer.

Interacting primers must contain a reverse complement of a run of a sequence (the run of complementarity can be either continuous or interspersed by noncomplementary base pairs). If the complement of a certain sequence stretch in one primer is present within the sequence of the other primer, there can be binding between different primer molecules. The bound molecules may remain as they are or be extended by the Taq polymerase. The extension is possible if the binding area happens to encompass the 3' end in one of the interacting molecules or by chance, the 3' ends of both molecules. The dotted line represents the extended portion of one molecule.

If the reverse complement is found within the same molecule, two identical molecules of primer can bind together and, in addition, a single primer molecule can form a hairpin. The hairpin conformation can undergo a transition to a state of two associated primer molecules and vice versa.

## PCR Troubleshooting

**X-legged hairpin**

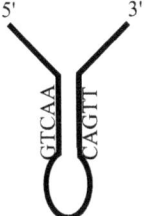

**Assymetric hairpin with protruding 3' end**

**Assymetric hairpin with protruding 5' end**

**Figure 2. Further examples of the hairpin structures.**

The hairpins can be "X-legged", or asymmetric. The asymetric hairpins can have protruding 5'ends or protruding 3' ends. The hairpin with protruding 5' end as well as the dimeric structure formed with two primer molecules of that kind could be extended by Taq polymerase.

It should be understood or be evident that the strand elongation carried out by Taq results in strengthening of the dimeric primer link or the hairpin, and this will result in approaching its $Tm$ to that of the $Tm$ of the correct bond between the primer and its target in the template DNA, or, in special rare cases, even raising the $Tm$ above that value.

Therefore, in primer selection there must be reasonable cutoff values for the stability of the primer associations and the hairpins expressed in any of the following terms: free energy, $T_m$, the maximal number of A-T and G-C bonds. For the structures that can be used by Taq polymerase the cutoff should be more strict than for the structures that cannot. The calculations of the cutoff values should be oriented to the lowest in the accepted range of the annealing temperatures, to the maximal in the range of the primer concentrations and to the standard ionic strength in the PCR buffer. As usual, there must be separate cutoff values for several nucleotides adjacent to the 3' end of the primer and for the whole of the primer. I am not aware of any detailed investigation into the cutoff question including calculations confirmed by extensive experimental data. When I select primers, I rely on my intuition.

**Advice:**
1. Raise the annealing temperature, try hot start, touchdown, reduce the annealing and elongation time, add helix destabilizers (see **HEDE**) and try any other means to enhance the reaction specificity (**EVRESP**).
2. Try **SBRD**.
3. See **NQWP**.
4. To resolve associated primers heat the primer stock solutions prior to use. Flash-cool them and keep them on ice. Assemble the PCR mix on ice. Minimize the time the tubes are kept at room temperature.

# SHORT PCR FRAGMENTS (UP TO 500BP) VERSUS LONG PCR FRAGMENTS (UP TO 4000BP)

## (SDVS)

| Table 2. Short PCR fragments versus long PCR fragments (SDVS) ||
|---|---|
| SHORT | LONG |
| • Usually robust PCR which does not require much optimization.<br><br>• Short extension times, down to several seconds.<br><br>• Very short denaturation times after a few (7-10) initial cycles with long denaturation times (as far as the template for the PCR is the added DNA rather than the accumulated PCR fragments).<br><br>• Add more primers, see **MPRE**. | • The template DNA must be completely denatured.<br><br>• Somewhat longer extension times due to longer distance to be filled in with nucleotides. Do further gradual extension of the DNA synthesis step in the last third of the process. In the last cycle abruptly extend the time for the DNA synthesis step.<br><br>• Somewhat longer denaturation times throughout the process or denaturation aided by cosolvents.<br><br>• Do rapid final cooling down to 8°C or lower.<br><br>• Add more dNTP's and perhaps more Taq.<br><br>• Take additional pains to ensure the purity of the mix including dNTP stock. |

## AVOIDING ACCIDENTS (ERRORS IN CALCULATIONS AND ERRORS IN PIPETTING)

### (*BLANK, LADD, SMEA, SPNO, TRIV*)

An experiment usually begins with complex arithmetics. Then you have to perform it in your imagination as a clearly understood succession of imaginary actions, beginning with the initial disposition of tubes on the bench. Then comes the actual implementation. This is usually pipetting. Put away the tube you have taken the liquid from, in order not to take from it a second time. Put the tube you have added the liquid to in a different position. In this way, "confirm" every action by moving the tubes. If you feel fidgety, excited, downspirited, fagged out, or if you are drunk, it is better to drink more than to begin this mind-on and hands-on work!

# CONCLUSIONS

## A FEW WORDS TO THE NOVICE

The downside of this and other compositions of the kind is the false impression of an enormous complexity of the method – the impression that could instil in you a deep mistrust in your own ability to cope with it. In fact, in most cases, the PCR is simple, especially if the fragment is short. Remember that nearly every advice or rule found in this or other manuals can be disregarded and yet there is a reason to hope for a successful PCR.

*"There are many other things in life than PCR."*

## A FEW WORDS TO A PCR ADEPT

There are many other things in life than PCR ... for example, isothermal DNA amplification. PCR may have a poor future because it requires a sophisticated and expensive apparatus – the thermocycler. Some methods of isothermal amplification that have been reported to be not inferior to PCR with regard to the sensitivity, specificity, and the feasibility of producing quantitative results require just a constant temperature of 37°C.

If PCR is to retain its place in the laboratory, it will certainly be in a different form to that described in this article. Agarose gel electrophoresis is already being superseded by the closed-tube detection of the PCR fragments by optical means. This relieves the operator from the plague of contamination. Lack of the quantitative effect is a major drawback of the conventional PCR method. In the real-time thermocyclers the sensitivity of the closed-tube detection is so greatly enhanced that there is the feasibility of detecting the specific PCR-fragment while the process of amplification is still in the logarithmic phase, making possible estimation of the PCR target copy number in the DNA sample.

Let me finish off with this conundrum: why, being so robust at incredibly high temperatures, is the Taq enzyme so fragile and vulnerable at room temperature?

# GLOSSARY

| Term | Page | Description |
|---|---|---|
| Action | 12 | The step that should be taken to alter a particular appearance ($qv$). |
| Adept |  | Someone considered highly proficient (in this book, refers to someone considered highly proficient in PCR). |
| ADNA | 19 | Additional DNA from impurities. |
| ADVI | 59 | How to avoid the generation of chimeras. |
| ALDR | 60 | Causes of allele dropout. |
| ALPS | 54 | Approaching the limit of the PCR sensitivity. |
| Appearance | 12 | What is actually seen in a gel: nothing, additional bands, smear, etc. |
| APPT | 20 | Suboptimal concentration of nucleotides exacerbates incomplete primer elongation. |
| AQWW | 24 | Qualities of the water used in PCR. |
| ATAS | 50 | Slow or delayed cooling in the end of the run may favor the formation of the multistrand molecular aggregates even when a PCR product comprises fully double-stranded molecules. |
| BBBB | 19 | Low primer, target, Taq, and nucleotide concentrations are to be favoured as these generally ensure cleaner product and lower background. |
| *BLANK* | 13 | Blank gel, no bands, with the possible exception of the bands at the bottom containing associated primers. |
| BTIP | 27 | Wood and wooden implements can contain PCR inhibitors. |
| Cause | 12 | What the appearance ($qv$) results from. |
| CDPR | 25 | Specific conditions for amplification of AT-rich, GC-rich and hairpin sequences. |
| CHSP | 32 | The active principle of the temperature regulation in thermocycler. |

Page numbers in **bold** type refer to pages containing the definition of the code or term. Codes for appearances are presented in *UPPERCASE BOLD ITALIC* codes for causes and actions in **UPPERCASE BOLD**.

| | | |
|---|---|---|
| **CQBA** | 42 | Denaturation of the PCR fragment is opposed by its renaturation at the annealing or synthesis steps at the latest cycles of PCR. |
| **DATP** | 35 | Counteracting temperature inactivation of the Taq enzyme by appropriate programming of the thermocycler. |
| **DELT** | 43 | Decrease the elongation temperature in amplifying AT-rich sequences. |
| **DGFTW** | 36 | A way to clean a thermocycler well to ensure a close contact between the tube and the well. |
| **DILU** | 23 | To counteract the action of PCR inhibitors dilute the template. |
| **EDNE** | 20 | Limits of dNTP concentration. Excessive dNTPs inhibit PCR. |
| **EQWT** | 42 | Increasing the elongation time is a way to counteract stable hairpins. |
| **EVAP** | 36 | Rapid evaporation of the tiny volume of liquid above the paraffin or directly from the tube. |
| **EVRESP** | 66 | Avoiding primer dimers by ways of raising the reaction specificity. |
| **FQWR** | 34 | The slow ramp means lingering about the temperature set for a step. |
| **FTSE** | 38 | Faulty target selection. |
| ***GHOS*** | 13 | A stained image of the well slowly migrating down the lane. |
| **GVPT** | 20-21 | Raising of primer actual concentration would not result in the raising of their effective concentration. |
| **HAIRP** | 42 | Hairpins in the internal (interprimer) region of a PCR fragment. |
| **HEDE** | 51 | Helix-destabilizers – the substances which promote DNA denaturation. |
| **HEST** | 43 | Within the AT-rich runs the nascent DNA strand dissociates from the template strand and its growth stops. |
| **HETE** | 60 | How to avoid generation of heterduplexes. |
| ***HMWG*** | 13 | DNA remains in the well and won't enter the gel. |

Page numbers in **bold** type refer to pages containing the definition of the code or term. Codes for appearances are presented in ***UPPERCASE BOLD ITALIC*** codes for causes and actions in **UPPERCASE BOLD**.

| | | |
|---|---|---|
| **HREN** | **57-58** | The significance of the difference in the salt concentration in the compared samples. |
| **HTHT** | 20 | Rules for handling tiny amounts of DNA. |
| **IFNO** | 60 | Extension of the DNA synthesis step in the last cycle followed by rapid final cooling could be helpful against heteroduplexes. |
| **IPEL** | 48 | Incomplete primer elongation or premature termination of DNA synthesis during the elongation step of a PCR cycle– implications and remedies. |
| **IPMC** | 29 | Spontaneous thawing of reagent solutions in a freezer. |
| **ITFM** | 40 | To denature the template DNA begin the run with minutes at 95°C or 15-20 seconds at 97°C. |
| **IUYU** | 27 | Causes of the presence of pyrophosphate in dNTP stocks and in the PCR mix. |
| **KKIM** | 30-31 | Conditions for storage of dNTPs and the primers. |
| **KLMN** | 23-24 | Verification of the integrity of the template DNA by means of electrophoresis. |
| **KPPT** | 21 | Consequences of suboptimal and excessive Taq concentration. |
| ***LADD*** | 13 | Ladder, one or several bands instead of, or in addition to, the required band; low molecular weight entities comprising primer aggregates at the very bottom of the gel are not considered as components of ladder. |
| **LGCR** | 41 | The longer or GC-rich PCR fragments require longer denaturation times or higher denaturation temperatures. |
| **Lore** | | Knowledge gained through tradition or anecdote. |
| **MDIS** | 41 | Degree of the template DNA fragmentation depends on the method of its isolation. |
| **MINI** | 40 | To facilitate denaturation of the template DNA reduce its concentration. |
| **MPRE** | 21 | Calculation of the primer concentration needed to amplify short PCR targets. |

Page numbers in **bold** type refer to pages containing the definition of the code or term. Codes for appearances are presented in ***UPPERCASE BOLD ITALIC*** codes for causes and actions in **UPPERCASE BOLD**.

| | | |
|---|---|---|
| NQWP | 25 | Primers may not be good in practice even if they are good in design. Try the nested PCR. |
| PATS | 25 | The computer design of a PCR reaction usually implies four steps. |
| *PBOI* | 13 | The proper band occupies improper position in the gel. |
| PCDS | 25 | Primer selection tips. |
| PPBA | 26 | Recognition of a difference between mispriming and the inefficient priming. Advice to avoid inefficient priming. |
| PQFD | 41 | Usefulness of helix-destabilizers in denaturation of long or GC-rich PCR fragments. |
| PRDI | 63 | Avoiding primer-dimers and primer multimers. |
| PRST | 21 | A gross change in the dNTP concentration would require a change in the concentration of $MgCl_2$. |
| PTTM | 55 | Conditions for enhancing the frequency of the primer and its annealing site encounters at the very low copy number of the PCR target; the same for the Taq polymerase and the 3'terminus of the annealed primer. |
| PYRO | 31 | One of the major products of deoxynucleotidetri phosphate degradation is pyrophosphate, known to be the inhibitor of any DNA polymerase. |
| READ | 21 | A change in the KCl-based buffer concentration or any other component of the PCR mix may require readjustment of the Mg2+ concentration. |
| REME | 49 | To prevent incomplete primer elongation increase the time for DNA synthesis in late PCR, especially in the last cycle. |
| RQAT | 34 | A slow transition from 94°C to 65°C in amplifying ATrich target would result in dissociation of a growing DNA chain from the template chain. |
| RQTE | 41 | Dependence of denaturation and renaturation kinetics of a DNA fragment on its length, time and concentration; use of restrictases to fragment the template DNA. |

Page numbers in **bold** type refer to pages containing the definition of the code or term. Codes for appearances are presented in ***UPPERCASE BOLD ITALIC*** codes for causes and actions in **UPPERCASE BOLD**.

| | | |
|---|---|---|
| **RQWN** | 19 | Accumulation of nonspecific products during reamplication. |
| **SBCYC** | 43 | A way to amplify difficult targets containing both AT-rich and GC-rich sequences. |
| **SBRD** | 15 | If one of the primers is supposed to be poor, do a PCR run with the good primer alone; then add the other primer and fresh Taq, and repeat the run. |
| **SDSI** | 28 | A tiny amount of SDS left over in the DNA prepared forPCR is sufficient to inhibit the Taq polymerase activity. |
| **SDVS** | 67 | A difference in conditions required for amplification of short and long PCR targets. |
| **SKCI** | 22 | Suboptimal KCl concentration in the PCR buffer or the whole of the PCR buffer. |
| ***SMEA*** | 13 | Smear, a smoothly stained area in the lane having no stepwise character of the ladder. |
| **SOLT** | 52 | Cosolvents aiding to straighten out DNA secondary structures and to prevent renaturation of the amplicon strands at the temperatures of the primer annealing. |
| **SPTPW** | 36 | Suboptimal performance of the thermocycler or its particular wells. |
| ***SPNO*** | 13 | Specific nonspecifics, unsatisfactory result may appear as a satisfactory one. A single band in the lane in the expectedposition in fact is not identical to the target that has had to be amplified. |
| **SQWM** | 40 | Efficient ways to denature template DNA prior to PCR in solutions of low salt concentration. |
| **TCBP** | 27 | Nonspecific products or PCR failure due to poor Taq. |
| **TDPB** | 29 | Conditions for the storage of the template DNA, primers and dNTPs. A definite way for the preparation of the negative controls. |
| **TEIN** | 27 | PCR inhibitors in the template DNA. |

Page numbers in **bold** type refer to pages containing the definition of the code or term. Codes for appearances are presented in ***UPPERCASE BOLD ITALIC*** codes for causes and actions in **UPPERCASE BOLD**.

| | | |
|---|---|---|
| **TKKT** | 37 | The tubes are poorly pressed down into the wells or they have got deformed. |
| **TMAC** | 52 | The use of tetramethylammonium chloride as a helix stabilizing agent to amplify AT- rich PCR targets. |
| **TOUC** | 55 | Implications of touchdown. |
| **Touchdown** | 55 | A full description of touchdown is beyond the scope of this book. Please consult another manual for the definition of this term. |
| *TRIV* | 61 | The trivial contamination or false contamination. |
| **TSPCR** | 42 | Two-step PCR with long primers is a way to counteract stable hairpins. |
| *UNTIDY* | 13 | Untidy appearance of the gel with stained blots, irregular patches, dots, etc. |
| **WQWH** | 32-33 | Gross changes in the PCR mix composition may put in disorder the temperature regulation in the thermocycler. |

Page numbers in **bold** type refer to pages containing the definition of the code or term. Codes for appearances are presented in ***UPPERCASE BOLD ITALIC*** codes for causes and actions in **UPPERCASE BOLD**.

# INDEX

| | |
|---|---|
| ADNA | **19,** 27 |
| ADVI | **59,** 50 |
| ALDR | **60,** 50 |
| ALPS | **54,** 18, 19 |
| APPT | **20,** 50 |
| AQWW | **24,** 27 |
| ATAS | **50,** 35 |
| AT-rich | **43,** 25, 34, 35, 52, 60 |
| BBBB | **19,** 20, 21, 21 |
| *BLANK* | 13 |
| BTIP | **27,** 14 |
| CDPR | **25,** 39 |
| CHSP | **32,** 34, 55 |
| Cosolvents | **51,** 28, 33, 40, 67 |
| CQBA | **42,** 43, 48, 50 |
| DATP | **35,** 41, 41, 43, 55 |
| DELT | **43,** 35 |
| DGFTW | **36,** 50 |
| DILU | **23,** 27 |
| EDNE | **20,** 50 |
| EQWT | **42,** 35 |
| EVAP | **36,** 46, 50 |
| EVRESP | **66,** 35, 61 |
| FQWR | **34,** 35, 43 |
| FTSE | **38,** 25 |
| GC-rich | **43,** 25, 41 |
| *GHOS* | 13, 58 |
| GVPT | **20-21,** 63 |
| Hairpins | **64-65,** 15, 25, 26, 42, 43, 60, 63, 66 |
| HAIRP | **42,** 43, 50, 52 |
| HEDE | **51,** 35, 40, 41, 42, 43, 66 |
| HEST | **43,** 50, 52 |
| HETE | **60,** 52 |
| *HMWG* | 13, 19, 19, 37, 57, 58 |
| HREN | **57-58,** 15 |

Page numbers in **bold** type refer to pages containing the definition of the code or term. Codes for appearances are presented in *UPPERCASE BOLD ITALIC* codes for causes and actions in **UPPERCASE BOLD**.

| | |
|---|---|
| HTHT | **20,** 27 |
| IFNO | **60,** 35 |
| Inhibitors | 23, 24, 27, 28, 48, 53, 62 |
| IPEL | **48,** 20, 21, 59, 60, 60 |
| IPMC | **29,** 24, 27 |
| ITFM | **40,** 35, 41 |
| IUYU | **27,** 31, 50 |
| KKIM | **30-31,** 24, 27 |
| KLMN | **23-24,** 41 |
| KPPT | **21,** 50 |
| *LADD* | **13** |
| LGCR | **41,** 35, 43 |
| Long PCR fragments | **67,** 20, 34, 35, 41, 48, 51, 55 |
| Low copy number of the PCR target | **18,** 54 |
| MDIS | **41,** 41, 55 |
| MINI | **40,** 41, 43 |
| MPRE | **21,** 50, 67 |
| NQWP | **25,** 50, 61, 66 |
| PATS | **25,** 42, 43, 43, 43 |
| *PBOI* | **13,** 46, 57, 57 |
| PCDS | **25,** 39 |
| PPBA | **26,** 35 |
| PQFD | **41,** 35, 43, 43, 50 |
| PRDI | **63,** 15, 25 |
| PRST | **21,** 20, 50 |
| PTTM | **55,** 35 |
| PYRO | **31,** 24 |
| READ | **21,** 50 |
| Renaturation | 41, 42, 43, 48, 51 |
| REME | **49,** 35 |
| RQAT | **34,** 35, 43, 50 |
| RQTE | **41,** 41, 43, 55 |
| RQWN | **19,** 14 |
| SBCYC | **43,** 35, 50 |
| SBRD | **15,** 25, 66 |

Page numbers in **bold** type refer to pages containing the definition of the code or term. Codes for appearances are presented in ***UPPERCASE BOLD ITALIC*** codes for causes and actions in **UPPERCASE BOLD**.

| | |
|---|---|
| SDSI | **28,** 53 |
| SDVS | **67,** 35 |
| SKCI | **22,** 40, 41, 42, 43, 43 |
| *SMEA* | 13 |
| SOLT | **52,** 35 |
| SPTPW | **36,** 50 |
| *SPNO* | **13,** 20, 20, 21, 25, 31, 31, 36, 37, 38, 52, 59, 59, 60, 61, 61, 63, 68 |
| SQWM | **40,** 41, 43 |
| TCBP | **27,** 50 |
| TDPB | **29,** 50 |
| TEIN | **27,** 23, 50 |
| TKKT | **37,** 50 |
| $T_m$ | 26, 35, 63, 65, 66 |
| TMAC | **52,** 35, 43 |
| TOUC | **55,** 35, 66 |
| *TRIV* | **61,** 68 |
| TSPCR | **42,** 35 |
| *UNTIDY* | **13,** 58 |
| WQWH | **32-33,** 50 |

Page numbers in **bold** type refer to pages containing the definition of the code or term. Codes for appearances are presented in *UPPERCASE BOLD ITALIC* codes for causes and actions in **UPPERCASE BOLD**.

# FURTHER READING

## Real-Time PCR: An Essential Guide
Publisher: Horizon Bioscience
Editors: Kirstin Edwards, Julie Logan and Nick Saunders
0-9545232-7-X  GB £105 or US $210.  http://www.horizonpress.com/rtpcr
Real-time PCR (RT-PCR) is a powerful and rapid technique for nucleic acid amplification. The accumulation of specific products in a reaction is monitored continuously during cycling. This is usually achieved by monitoring changes in fluorescence within the PCR tube. This essential manual presents a comprehensive guide to the most appropriate and up-to-date technologies and applications as well as providing an overview of the theory of this important technique. Written by recognized experts in the field this timely and authoritative volume is an essential requirement for all laboratories using PCR.  Topics covered include: Real-time PCR instruments and probe chemistries, set-up, controls and validation, quantitative real-time PCR, analysis of mRNA expression, mutation detection, NASBA, application in clinical microbiology and diagnosis of infection.

## DNA Amplification:
## Current Technologies and Applications
Publisher: Horizon Bioscience
Editors: Vadim V. Demidov and Natalia E. Broude
0-9545232-9-6   GB £105 or US $210.   http://www.horizonpress.com/dna
DNA amplification is the cornerstone of modern biotechnology and it is also a key procedure in numerous basic studies involving DNA and other biomolecules. The emergence of alternative methodologies to PCR has significantly widened the range of approaches for DNA amplification and dramatically improved the technological abilities of basic and applied researchers in various fields of life sciences. Whereas most books on DNA amplification focus on PCR-based technologies, this volume presents a wider range of methods to amplify DNA with an emphasis on their diverse applications.  The book covers both well-established and newly-developed protocols including ligation-based thermocycling approaches, real-time PCR and other new PCR developments, plus several powerful non-PCR isothermal DNA amplification techniques, for example: real-time strand displacement amplification (SDA), rolling-circle amplification (RCA) and multiple-displacement amplification (MDA). An entire section is devoted to a group of enzymes, both natural and engineered, which are employed for DNA amplification and related purposes. In addition, the use of DNA amplification in the detection of non-DNA analytes is presented. Written and edited by leading experts in the field, this book serves as a practical tool and an invaluable reference source for a broad audience of academic researchers and industry biotechnologists who use DNA amplification techniques.

## The Gateway to PCR
A major web resource for information and links on all aspects of the Polymerase Chain Reaction. PCR has revolutionised molecular biology over the last ten years. This web site aims to provide scientists with a comprehensive directory for the polymerase chain reaction, DNA amplification and PCR related techniques and protocols.

### http://www.horizonpress.com/pcr

www.ingramcontent.com/pod-product-compliance
Ingram Content Group UK Ltd.
Pitfield, Milton Keynes, MK11 3LW, UK
UKHW021322180426
11947UKWH00015B/1377